Common Core State Standards for High School Math:
What Every Math Teacher Should Know

GEOMETRY

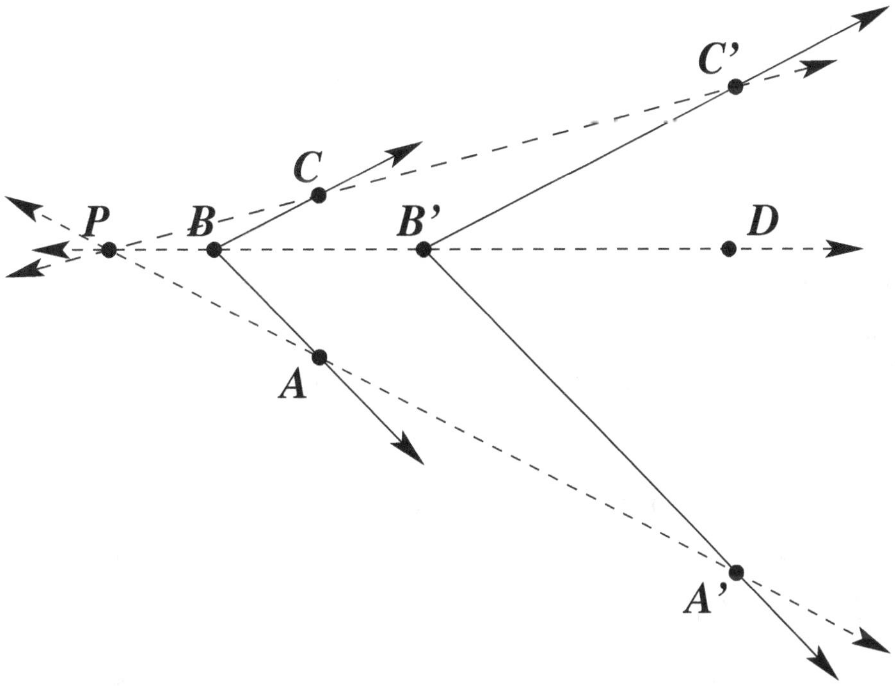

FOREWORD

Thank you for choosing this book, the first in what I hope will be a series explaining the Common Core State Standards in Mathematics (CCSSM) to a wide audience. The subtitle of the series "What Every Math Teacher Should Know" is perhaps a bit over the top, but it is true that the CCSSM have the potential to alter the K-12 mathematics education landscape radically. And so, I offer this book as a tool for teachers to help them understand *WHY* the CCSSM were written the way they were written, at least from my perspective.

Who am I, then? I'm a professional mathematician who has long been interested and engaged with the teaching of mathematics, from K-12 to University levels. I've helped out with many K-12 professional development activities, and I've even written a few study guides to help California's math teachers with their required standardized tests. In those guides, I followed a basic approach: start with the standards (verbatim) that the teachers need to know and explain them to the best of my ability. I have tried a similar approach here[1]. In the following pages, I list each standard and substandard verbatim, and follow them with a serious of Socratic questions, designed to lead the reader through the vocabulary and the rationale behind each standard. I even demonstrate a few examples and solve a few sample problems along the way. Ideally, the reader will begin to understand *HOW* the CCSSM view of Geometry was devised. Understanding the bigger picture will help to understand the mathematics.

The book follows the same structure as the CCSSM Geometry Standards, which are organized into six sections with several subsections. The Table of Contents can be found on the next page.

Best of luck as you read the book, which contains over 100 figures, mostly drawn by me using xfig[2], my favorite drawing program, but I also used Geogebra[3] on occasion. May you find in these pages what you are looking for! –CG

---

[1]Indeed, some of this material is from the California guides, though most material is new.
[2]http://xfig.org/
[3]http://www.geogebra.org/cms/

CONTENTS

GEOMETRY STANDARDS

CONGRUENCE

- Experiment with transformations in the plane

1. Know precise definitions of angle, circle, perpendicular line, parallel line, and line segment, based on the undefined notions of point, line, distance along a line, and distance around a circular arc.

   - First, a disclaimer. While I am trying to be very precise and exact in what follows, this text is not intended to be a substitute for a good geometry textbook. In particular, it is difficult for this book to be mathematically constructed, building only on previously explained material, without going into a lot more detail. Please consult a good geometry text for further information.
   - Why are point, line, distance along a line, and distance around a circular arc undefined? How are they notated?
     Any definition has to rely on other words. Those words, then, must also be defined. So, there comes a point when one must either accept the fact that there will be circular definitions (e.g., when a word appears in its own definition) or else there must be terms that are not defined, that everyone accepts as commonly held notions or beliefs. In geometry, "point" and "line" are examples of those undefined terms. Most people can understand and largely agree on the abstract ideas of a point and a line even though those terms are technically not defined. We will denote a point by a capital letter, like P, and a line by listing two of the points on it and using a line symbol, like $\overleftrightarrow{AB}$.
     Also, notions of distance are highly subjective. The length of a segment of a line (or distance along a line) depends on what the unit of distance is, and the scale. So, while you may be able to make comparisons (longer, shorter, larger, smaller, etc.), an absolute measure of distance is not defined. Similarly, distance around a circular arc is an undefined concept, though we will assign it a sense of measure later.
   - What are the precise definitions of (and notations for) angle, circle, perpendicular line, parallel line, and line segment?
     * Before we can define an angle, we need to define a "ray." A ray is the set of all points on a line that lie on one side of a given point, called the endpoint of

the ray. If the ray is contained in line $\overleftrightarrow{AB}$, has an endpoint at A, and includes point B, then we denote the ray as $\overrightarrow{AB}$.

* An "angle" is the geometric object formed by two rays emanating from a single point. The common endpoint is called the "vertex" of the angle, the two rays are called the "sides" of the angle, and the region between the sides is called the "interior" of the angle. If the rays are $\overrightarrow{AB}$ and $\overrightarrow{AC}$, then we call the angle $\angle BAC$, or even just $\angle A$ is there is no possibility for confusion.

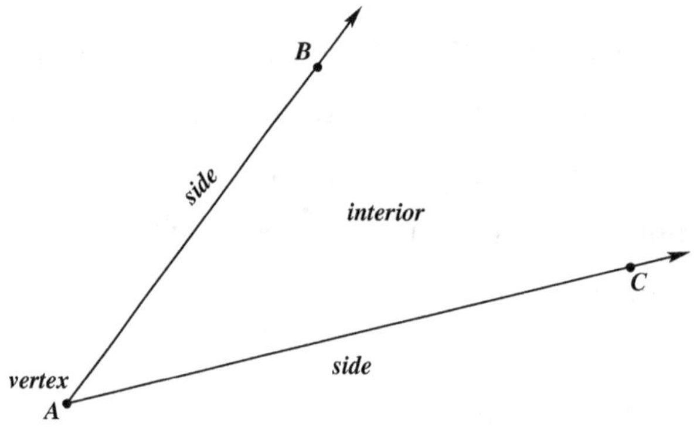

Some other terms involving angles: $\angle BAC$ is a "straight" angle if points A, B, and C are collinear, with A between B and C. We will define angle measure so that a straight angle measures 180 degrees. Two angles are "adjacent" if they share a common vertex and a common side but have disjoint interiors. Adjacent angles are "supplementary" if, when taken together, they form a straight angle. Two congruent supplementary angles are called "right" angles. (A precise and independent definition of "congruent" will come later.) We have already assigned a sense of angle measure to straight angles. We could say that a right angle has a measure of 90 degrees. Using this standard, we occasionally refer to the measure of $\angle BAC$, denoted $m\angle BAC$. This is defined to be a number between 0 and 180 degrees, and to obey some rules regarding angle addition and congruence. Consult a good geometry text for more information.

* A "circle" is the geometric object formed by the set of all points that lie a given distance (called the "radius") from a given point (called the "center"). If the center is point P, then the circle is often called "circle P." The radius is not explicitly stated unless there is some possibility for confusion (e.g., when there are two circles centered at P with different radii).

* Two lines are "perpendicular" if they intersect at a right angle, or, put another

way, if they intersect to form four congruent angles.

* Two lines are "parallel" if they do not intersect. (We are assuming here the geometry of the plane. A three-dimensional version might be that two lines are parallel if they lie on one plane but do not intersect.) Notice that this is a very geometrical definition. We will not talk about "slope" in geometry until we get to coordinate geometry, where algebra is applied to geometry.
* A "line segment" consists of all the points on a line that lie between two given points, called the "endpoints" of the segment. A line segment is denoted by its endpoints, like $\overline{AB}$. We will often assign a sense of measure to lines, at which point AB will denote the length of $\overline{AB}$.

- Why are these precise definitions?

  These definitions are precise because they depend only on terms that have been previously defined, or that are commonly accepted as undefined, like "point" or "line." Also, there should be no ambiguity as to whether a geometrical object meets the definition or not. Further, definitions are "biconditional" or if-and-only-if. That means that everything that meets the definition of a circle is a circle, for example. But it also means that if something does not meet the definition of a circle, then it is not a circle.

  Definitions need not be exclusive, though. For example, a square also meets the definition of a rectangle. So a square is both a square and a rectangle (and a rhombus, a parallelogram, and a quadrilateral).

2. Represent transformations in the plane using, e.g., transparencies and geometry software; describe transformations as functions that take points in the plane as inputs and give other points as outputs. Compare transformations that preserve distance and angle to those that do not (e.g., translation versus horizontal stretch).

- What is a transformation in the plane?

  A transformation in the plane is a function that maps points on the plane to other points on the plane. For instance, a function which moves every point on the plane 2 units to the right is a transformation (often called a translation, or a horizontal shift). In general, transformations don't have to be smooth or even continuous, but the ones we study will be reasonably well behaved.

- Which transformations preserve distances and angles? Why are they called "isometries?"

The most common transformations that preserve distances and angles are: rotations, reflections in a line, and translations. The word "isometry" comes from Greek words meaning "equal measure." That's why the word isometry is used to describe a transformation that has the additional property that all distances between points are preserved. So, if we call our isometry F, then the distance between points A and B is equal to the distance between $F(A)$ and $F(B)$. A common notation is to use primes for the transformed points. So, under an isometry, $A'B' = AB$. Because isometries preserve lengths, they also preserve any geometric property that follows from lengths, such as congruence of triangles, and thus of angles, polygons, and circles, too, as we will see. Areas and volumes are also preserved by isometries.

– Which transformations are not isometries?

Some common types are dilations centered at a point (e.g., all the points doubling their distance away from a fixed point), contractions centered at a point (e.g., all points moving a third of the distance toward a fixed point), dilations away from a line (e.g., all points moving twice their distance away from a fixed line), and contractions toward a line (e.g., all points moving half the distance toward a fixed line). There are many other transformations that do not preserve distances.

3. Given a rectangle, parallelogram, trapezoid, or regular polygon, describe the rotations and reflections that carry it onto itself.

– What is a "symmetry?" Why are they important?

A symmetry of a geometric figure is an isometry that doesn't really change the figure. For example, if you rotate a square counter-clockwise around its center by 90 degrees, you obtain a square that looks exactly like the original square. Thus, counter-clockwise rotation by 90 degrees is a symmetry of the square.

Symmetries are important because they bring about a deeper understanding of geometry by allowing algebraic properties into the picture in a few ways. First, one can compose two symmetries together to obtain another one. For example, counter-clockwise rotation by 90 degrees followed by another counter-clockwise rotation by 90 degrees gives a counter-clockwise rotation by 180 degrees, which is another symmetry of the square. This set of symmetries serves as a launching point for the rich and beautiful topic of groups in abstract algebra. Second, knowing the symmetries of an object can give you information about the object.

For example, knowing that a square has four rotational symmetries means that its four interior angles have to be congruent to each other, which then means that each must be a right angle. Physicists are experts at using symmetry to come up with their theories.

- How can we describe the rotations and reflections that carry each of these figures onto itself?

  Let's start with a non-square rectangle. How many rotational symmetries does it have? The answer is two: one is a counter-clockwise 180-degree rotation and one is a counter-clockwise 360-degree rotation. (While it's true that the 360-degree rotation – the IDENTITY transformation – is the same as doing nothing, it's also true that in terms of symmetries, that is as important as zero is to addition. So we count it, but we will call it a 0-degree rotation from now on.) What about a 540-degree counter-clockwise rotation? Or a 180-degree rotation clockwise? Well, it turns out that these are each the same transformation as the 180-degree counter-clockwise rotation, in that, as a function, each maps the points on the plane to the same final points. So we do not count these transformations as being distinct.

  What about reflections that preserve the rectangle? If we think about lines that could be mirrors, there are two lines of reflection that divide the rectangle in half. Each line starts at a midpoint of a side of the rectangle and travels to the midpoint of the opposite side (below).

  Also in the picture below, we have a non-rectangular parallelogram. This figure only has two symmetries, and they are both rotational: by 0 and by 180 degrees.

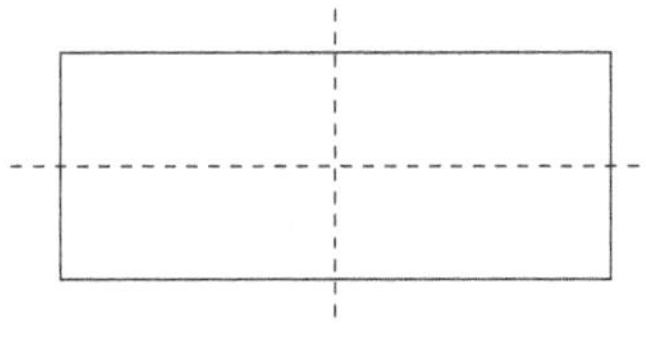 

The case of a trapezoid is interesting because in general, there is only the one symmetry every figure has: a rotation by 0 degrees. But if the trapezoid happens to be an isosceles trapezoid (in which the legs are congruent), then there is an additional reflection symmetry along the line that joins the midpoints of the trapezoid's bases.

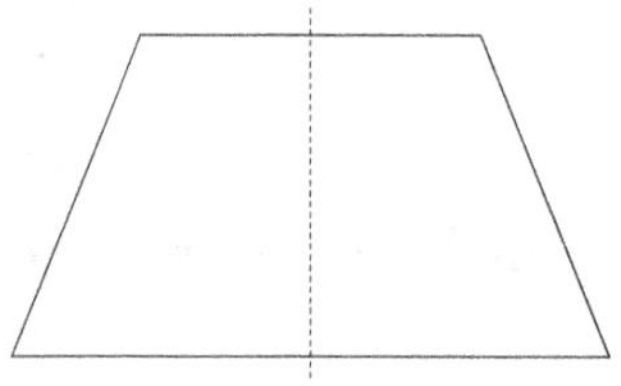

Regular polygons are much more symmetric. Consider the equilateral triangle and square below. (We have already discussed the rotational symmetries of the square, which we will revisit here.)

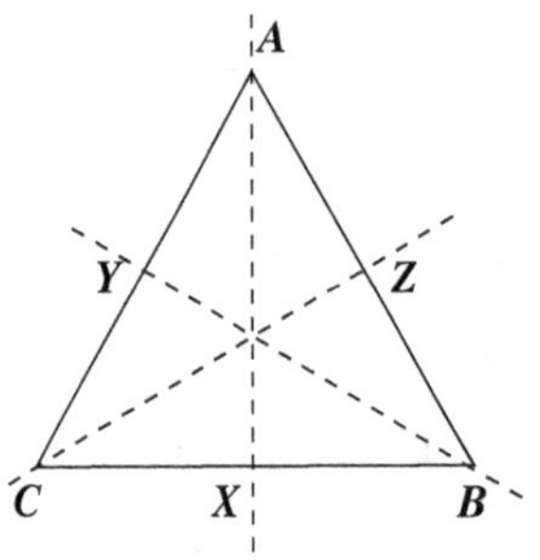

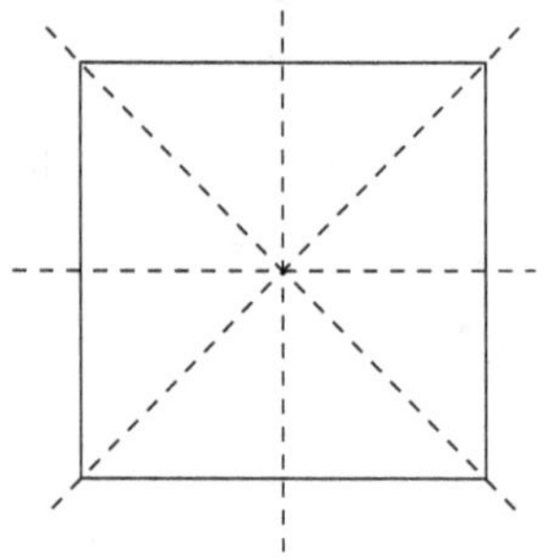

In the triangle above, there are six symmetries: three reflections in the dashed lines $\overleftrightarrow{AX}$, $\overleftrightarrow{BY}$, and $\overleftrightarrow{CZ}$; and rotations around the center by 120 degrees, 240 degrees, and 0 degrees.

In the square, there are eight symmetries: four reflections in the dashed lines, and four rotations, by 90, 180, 270, and 0 degrees.

There is a pattern developing here. You can probably guess that a regular pentagon has 10 symmetries: five rotations and five reflections. Also, a regular hexagon has 12 symmetries: six rotations and six reflections, etc. In general, a regular n-gon will have $2n$ symmetries: n rotations (multiples of $\frac{360}{n}$ degrees) and n reflections (through each line of symmetry through the center). [There is a subtle distinction in describing these lines of reflection depending on when n is even or odd. Can you figure it out?]

4. Develop definitions of rotations, reflections, and translations in terms of angles, circles, perpendicular lines, parallel lines, and line segments.

   – What information is necessary to describe a rotation? . . . a reflection? . . . a translation?

     * To describe a rotation, you need to state the center of rotation and the angle of rotation.

     * To describe a reflection, you need to state a line of reflection.

  * To describe a translation, you need to give a vector of translation; that is, you need to state a direction in which to translate and how far to move in that direction.

- How can we turn this information into precise definitions involving geometric objects?
  * Rotations require a center of rotation and an angle of rotation (counter-clockwise) around that center. Let's call the center of rotation C and say that we are rotating d degrees counter-clockwise around C. So we need to describe what happens to an arbitrary point P on the plane after it goes through this rotation of d degrees, centered at C. Let P' denote the image of P.

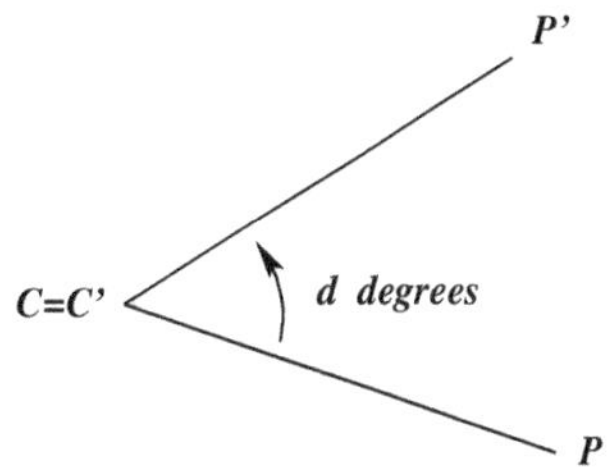

First, notice that C doesn't move: $C = C'$. If P is not C, then two properties determine P'. First, $m\angle PCP' = d$. Second, $PC = P'C$. So we have described this transformation in terms of angle measure and segment length. That is, the image of P under this rotation is the point P' that satisfies the two properties listed here.

How do we know that there is only one such point? Well, if we consider only the angle condition, then P' could be any point on a ray emanating from C. But the second condition means that P' lies on a circle of radius PC centered at C. There is only one place where the ray and the circle intersect. That point must be P'.

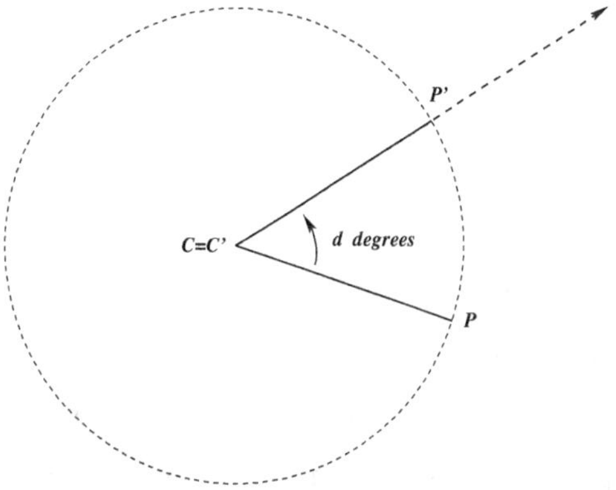

* What about a reflection? Well, the line of reflection (mirror line) is the most important feature. Let m be the line of reflection. What happens to an arbitrary point P after reflection in line m? Where is P'?

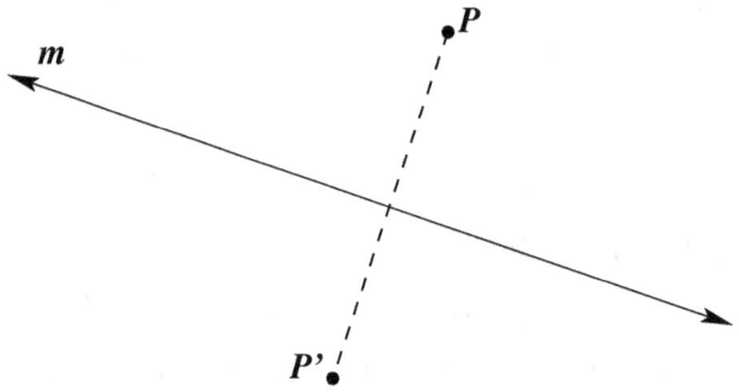

First, if P lies on line m, then the reflection doesn't do anything: $P = P'$. But if P is on one side of m, then P' is on the other. What can we say about the relationship between line m and segment $\overline{PP'}$? Line m must be the perpendicular bisector of $\overline{PP'}$. That is, the image of P under a reflection in line m is the point P' with the property that line m is the perpendicular bisector of $\overline{PP'}$.

How do we know that only one point has this property? Well, let's draw the unique line perpendicular to m that contains P. Let A be the point on m so that $\overleftrightarrow{PA}$ is perpendicular to m. Then we need $P'A = PA$. That is, P' must lie on line $\overleftrightarrow{PA}$ and on a circle of radius PA centered at A. Other than P, there is only one other point meeting those criteria. That point must be P'.

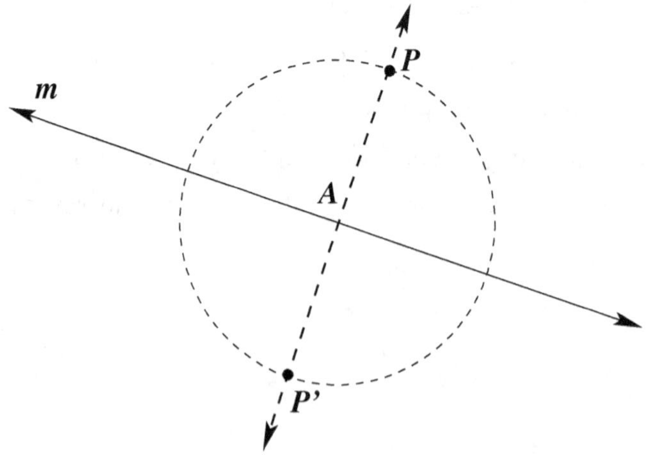

* For a translation, we need to state a direction in which to translate, and a distance to move in that direction. Or, put another way, if we know a point and its image, say A and A', then we can determine the image of any other point under this translation. It's easy to draw P', the point where the

translation would send P, but it's more difficult to explain it geometrically.

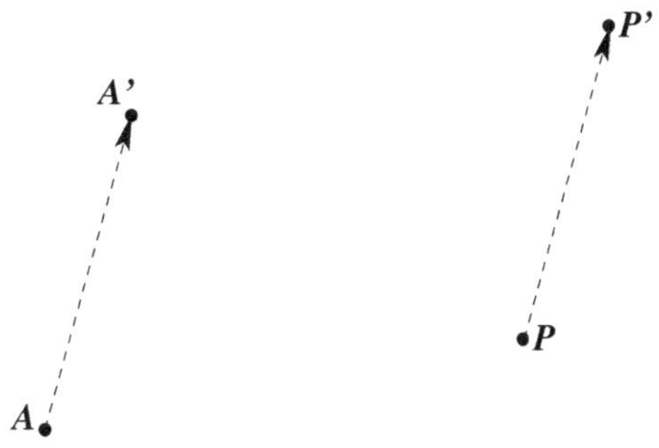

First, consider the case where P is located on $\overleftrightarrow{AA'}$. Then P' would also be on $\overleftrightarrow{AA'}$ and we would have $PP' = AA'$ and $AP = A'P'$. We know this point is unique: there are only two possible points P' on $\overleftrightarrow{AA'}$ satisfying $PP' = AA'$. (See picture, below.) The second equation forces the ray $\overrightarrow{PP'}$ to be pointing in the same direction as the ray $\overrightarrow{AA'}$.

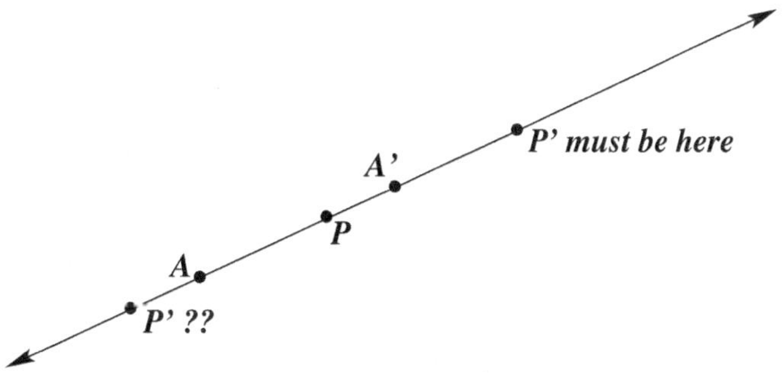

Second, consider the case when P is not located on $\overleftrightarrow{AA'}$. We still require $PP' = AA'$ and $AP = A'P'$, but this may not be sufficient to determine P' uniquely. For instance, just using these two criteria, we could have $APA'P'$ a square, and that's not what we want for a translation.

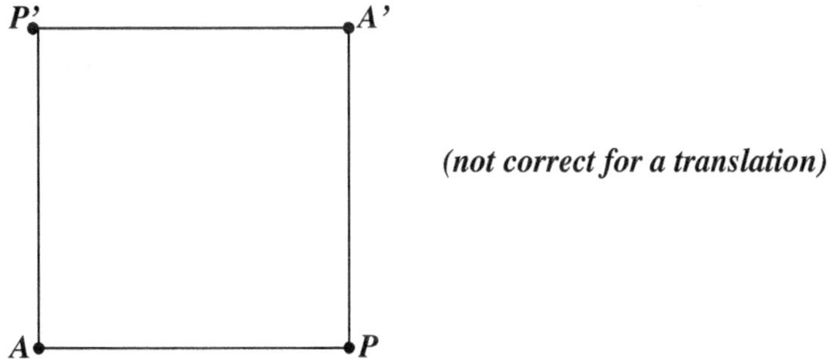

So we need another condition requiring $PAA'P'$ to be a parallelogram with vertices **in that order**: namely, that $\angle PAA'$ and $\angle AA'P'$ need to be supplementary angles.

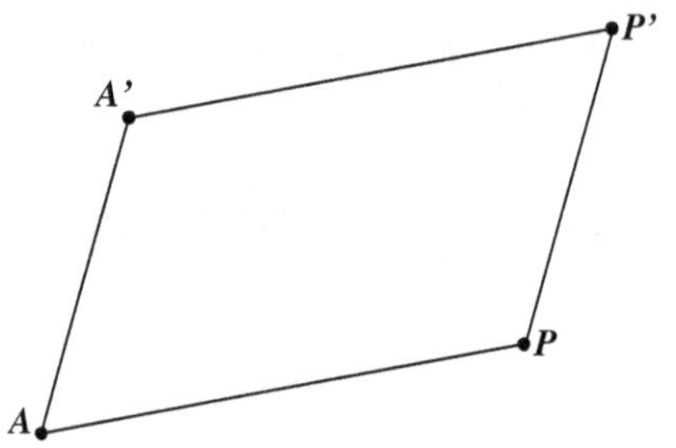

5. Given a geometric figure and a rotation, reflection, or translation, draw the transformed figure using, e.g., graph paper, tracing paper, or geometry software. Specify a sequence of transformations that will carry a given figure onto another.

   – How do you use geometry software?

     Effectively using geometry software goes beyond the scope of this book. Two popular software packages include Geometer's Sketchpad (Key Curriculum, information at http://www.keycurriculum.com/products/sketchpad) and GeoGebra (free, open source software, available at http://www.geogebra.org/cms/en). There are many other software packages available.

   – How do you draw a figure transformed by a rotation? Which media work best?

     Using graph paper implies the use of coordinates. Using what we know so far, it would be difficult to rotate a figure using graph paper unless the angle of rotation is some multiple of 90 degrees. See table below for the image of the point (a, b) under a counter-clockwise (ccw) rotation centered at the origin.

If the ccw rotation around the origin measures:	then the image of (a, b) is:
90 degrees	$(-b, a)$
180 degrees	$(-a, -b)$
270 degrees	$(b, -a)$

     To draw a rotated figure, then, rotate the important points first. For example, suppose you are rotating a triangle around the origin ccw 90 degrees and its vertices are at $(0, 5)$, $(-1, -2)$, and $(4, 3)$. Then the transformed triangle will have vertices at $(-5, 0)$, $(2, -1)$, and $(-3, 4)$. Connect the vertices to draw the rotated triangle.

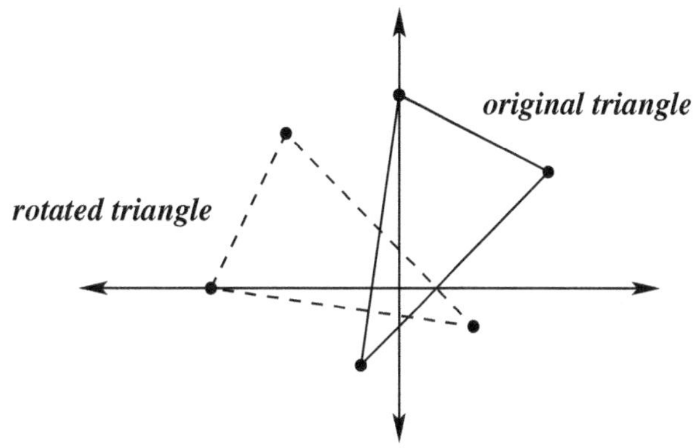

However, tracing paper might be easier in that it allows you to rotate any number of degrees, using any point as a center. Simply draw the figure, rotate the tracing paper around the center of rotation, and draw the new figure. Transparencies would work well too.

– How do you draw a figure transformed by a reflection? Which media work best?
Tracing paper would probably work best for a reflection, because the figure could be reflected in any line. Simply fold the paper along the line of reflection and then trace the image of the figure on the paper. Unfold the paper and you should see the original image and its reflection in the line.
Using transparencies would also work. For example, say you wanted to draw a triangle ($\triangle ABC$) and its image under reflection across one of its sides ($\overline{AB}$). Draw the same triangle on two different transparencies. Then flip one sheet over and lay it back down so that the two A vertices are on top of each other and the two B vertices are on top of each other. Then you are looking at the reflection of $\triangle ABC$ in side $\overline{AB}$.

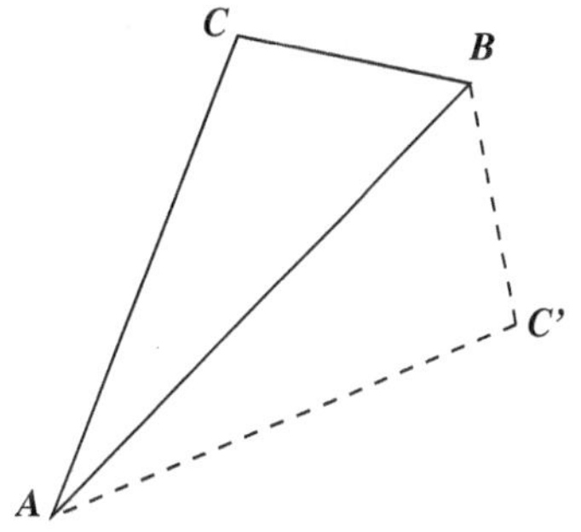

Reflections using graph paper (and coordinates) are a little trickier, unless the line of reflection is one of the following: $y = x$, $y = -x$, the x-axis, or the y-axis. See the following table for the image of (a, b) under reflections in these four lines.

If the line of reflection is:	then the image of (a, b) is:
$y = x$	(b, a)
$y = -x$	$(-b, -a)$
$y = 0$ (x-axis)	$(a, -b)$
$x = 0$ (y-axis)	$(-a, b)$

As before, simply draw the images of important points first and then determine the figure. For example, suppose you are reflecting the earlier triangle across the y-axis. (Recall that its vertices were at $(0, 5)$, $(-1, -2)$, and $(4, 3)$.) Then the reflected triangle will have vertices at $(0, 5)$, $(1, -2)$, and $(-4, 3)$. Now simply connect the vertices to draw the reflected triangle. Notice that the first vertex didn't change. That's because it is on the y-axis, and therefore its reflection is also on the y-axis.

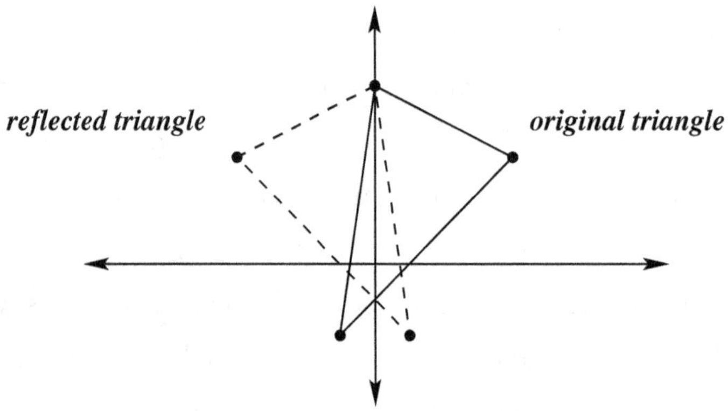

- How do you draw a figure transformed by a translation? Which media work best? Again, with tracing paper and transparencies, the situation is relatively easy. Simply draw the original figure on the tracing paper or transparency and slide the drawing to its new translated position. Then draw the translated figure there. With graph paper and coordinates, translations are also pretty straightforward, assuming we know the coordinates to which the origin moves under the translation. If the origin moves to the point (h, k), then the image of any point (a, b) under this translation will be $(a + h, b + k)$.

  As above, simply draw the images of important points first and then determine the figure. For example, suppose you are translating the earlier triangle. (Recall that its vertices were at $(0, 5)$, $(-1, -2)$, and $(4, 3)$.) If the origin moves to the point $(3, -1)$ under this translation, then the translated triangle will have vertices at $(3, 4)$, $(2, -3)$, and $(7, 2)$. Now simply connect the vertices to draw the translated triangle. Notice that there are no fixed points under a translation.

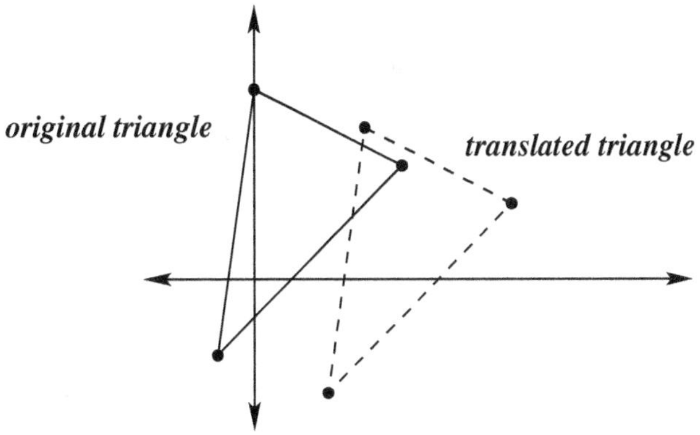

- What is a glide reflection and how does it work?

  As an example of a symmetry that is composed of two smaller symmetries, a glide reflection is made up of a translation and a reflection. The quintessential example of something exhibiting glide reflection symmetry is if you were to look at your footprints after walking on the beach. From one step to the next, each footprint is a reflection of the previous footprint, but the reflection is not a mirror image in some line. Instead, the right footprints are translated reflections of the left ones, and vice versa.

- How can you tell what sequence of transformations will take one figure to another?

  We just saw that a glide reflection is made up of a reflection and a translation. But what about more complicated figures?

  The good news is that, in the plane, you could actually make up any isometry with a sequence of reflections. For instance, to rotate 180 degrees around the origin, you could also reflect in the x-axis, and then reflect in the y-axis. The net result is the same as rotation by 180 degrees around the origin.

  More good news is that there is more than one route to get from one figure to another. But there are some "rules" that can help you figure out what types of transformations you are looking for.

  (a) Is the transformed figure the same size as the original figure? If so, then you only need to look at isometries. If not, then you will need to consider a dilation (if the transformed figure is bigger than the original) or a contraction (if the transformed figure is smaller than the original) at some point.

  (b) Can you determine if the transformed figure has the same orientation as the original figure? For instance, if you are looking at a non-isosceles right triangle, then traveling from the short leg to the long leg to the hypotenuse involves moving either clockwise or counter-clockwise around the triangle. Is the answer the same on the transformed triangle? If so, then you can reach the transformed figure without using a reflection. If not, then there is a reflection involved.

As I said, there are multiple correct solutions to the question of what transformations are involved in moving one figure to another. But here's a general idea involving a situation when the transformed figure is congruent to the original figure and doesn't require a reflection. Identify one point on the original (A) and its corresponding point on the transformed figure (A').

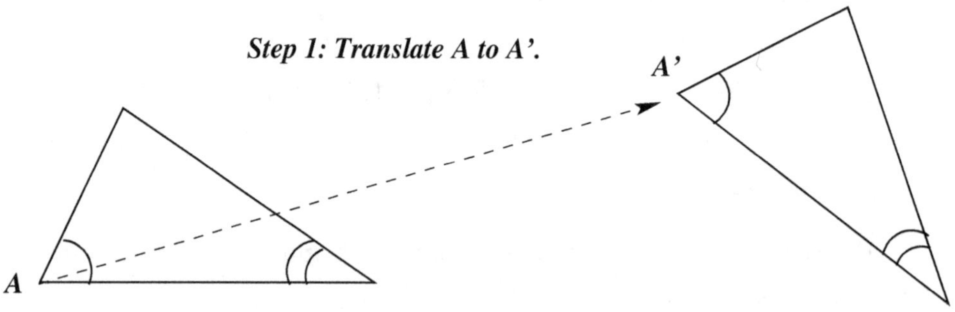

Step one: translate the original figure so that A moves to A'.

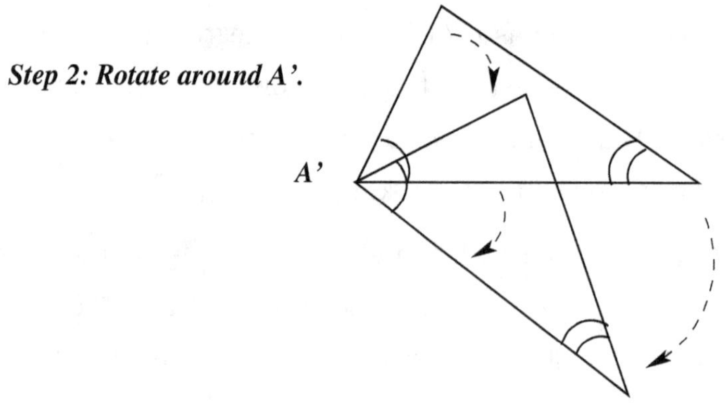

Step two: rotate around point A' until the original figure lines up with its image.

The triangles coincide.

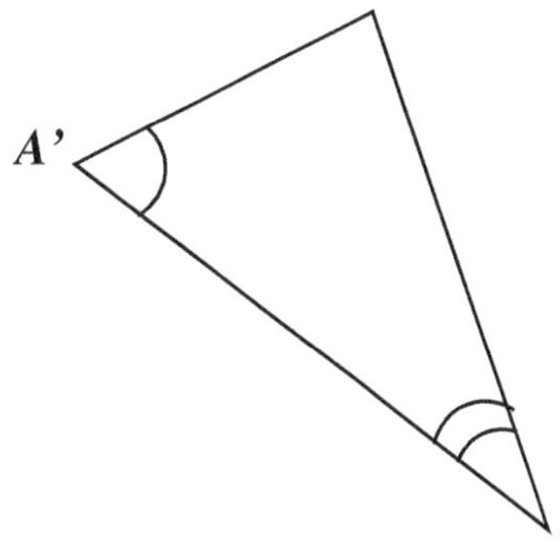

Thus this kind of problem can be done with two isometry transformations: a translation followed by a rotation. If the orientation of the initial object doesn't match the orientation of the final object, then you would also need to use a reflection.

- UNDERSTAND CONGRUENCE IN TERMS OF RIGID MOTIONS

6. Use geometric descriptions of rigid motions to transform figures and to predict the effect of a given rigid motion on a given figure; given two figures, use the definition of congruence in terms of rigid motions to decide if they are congruent.

   – What is another term for rigid motion?
   We have been using the term "isometry" in place of "rigid motion." Both mean the same thing, namely, a transformation of the plane that preserves distances. A consequence of this is that angles are also preserved under a rigid motion. It might help to think of geometric figures as being made out of a rigid material, e.g. a wooden triangle. Then a rigid motion can be achieved by simply moving the wooden triangle around: translating, rotating, and reflecting it.

   – How do you predict the effect of a given rigid motion on a given figure?
   One way is to continue to think of the figure as being made of a rigid material, like wood. Then imagine what would happen if you moved that wooden figure according to the particular function. So, translating a triangle could literally mean picking up the wooden triangle and sliding it to a new location without changing its orientation (i.e. without rotating or reflecting it). Reflecting the triangle in one of its sides could mean holding the two endpoints of that side in your hands and flipping the triangle over, but keeping your hands where they are. Rotating the triangle around one of its vertices could mean holding that vertex

in place with your finger and rotating the rest of the triangle around that fixed point.

- What is the definition of congruence in terms of rigid motions?

  Many people have a common-sense notion of what it means for two figures to be congruent: the figures need to be "the same." But a precise definition can be tricky. The Common Core State Standards are choosing to use a definition of congruence based in transformations, namely, that two figures are congruent if one can be transformed to the other via a sequence of rigid motions. In other words, two figures are congruent if you can perform a sequence of isometries (rigid motions) on one of them so that it coincides with the other.

- How do you use the definition of congruence in terms of rigid motions to decide if two figures are congruent?

  The short answer to this question is that you have to determine whether there is a series of isometries that would cause one figure ultimately to coincide with the other one. Since we are only allowed to use rigid motions, we know that if the two figures aren't the same size, then there is no way for them to be congruent.

7. Use the definition of congruence in terms of rigid motions to show that two triangles are congruent if and only if corresponding pairs of sides and corresponding pairs of angles are congruent.

- What do they mean by "if and only if"?

  Many proofs in math are conditional (or "if-then") statements, comprised of a hypothesis (the "if" part) and a conclusion (the "then" part). For example: if $x = 2$ then $x^2 = 4$. This is a true statement. You prove it by assuming the hypothesis is true (i.e. that $x = 2$) and then deducing the conclusion from that (i.e. $2^2 = 4$ is true).

  The converse of a conditional statement switches the role of the hypothesis and conclusion, giving you a different statement. The converse of the above example is: if $x^2 = 4$, then $x = 2$. We know this is a different statement, because it's not true. If we start by assuming $x^2 = 4$, we cannot deduce that x has to be 2. It is possible that $x = -2$ as well. So a converse statement is different than the original conditional statement. Moreover, the "truth value" of each of these statements is independent of the other.

  Biconditionals ("if and only if") statements are those statements for which the original and the converse are both true. For example, if $x = 2$, then $x + 2 = 4$.

This is true. Its converse, if $x+2=4$, then $x=2$, is also true. Therefore we could say, $x=2$ if and only if $x+2=4$. An if-and-only-if statement is really two conditional statements. So to justify such a statement, we really must justify two separate statements.

- How can you show that two triangles are congruent if their corresponding pairs of sides and corresponding pairs of angles are congruent?

  We assume that the corresponding pairs of sides and corresponding pairs of angles are congruent. Then that means that we could transform one of the triangles so that it coincides with the other. (For an example, see the last paragraph of #5, just above.) To make it easier, let's say that $\triangle ABC$ and $\triangle A'B'C'$ meet the hypotheses, and that each vertex corresponds with its primed counterpart: A with A', etc. Because the triangles correspond, there must be some sequence of rigid motions, possibly made up of translations, rotations, and reflections, which takes $\triangle ABC$ to $\triangle A'B'C'$. By the definition of congruence given above, this means that the two triangles are congruent.

- How can you show that the corresponding pairs of sides and corresponding pairs of angles of two triangles are congruent if the triangles are congruent?

  This is the converse of the previous statement. Here, we assume that two triangles are congruent. That means, by definition, that there is a sequence of rigid motions that takes one triangle to the other. Suppose we perform those rigid motions. What does the final picture look like?

  It looks like one triangle because the one triangle will be lying exactly on top of the other one. They coincide. Therefore, each side (respectively, each angle) of the first triangle corresponds to a side (respectively, an angle) of the second triangle. Corresponding pairs of sides and corresponding pairs of angles are congruent.

8. Explain how the criteria for triangle congruence (ASA, SAS, and SSS) follow from the definition of congruence in terms of rigid motions.

   - What is meant by ASA, SAS, and SSS?

     * We will begin with SSS. This stands for side-side-side, and it is a condition for triangle congruence. Specifically, it means that if you can match up two triangles in such a way that their three pairs of corresponding sides are congruent, then the triangles are congruent. We saw in #7, above, that two triangles are congruent if their corresponding pairs of sides and corresponding pairs of

angles are congruent. That's really six pairs of geometric objects. The SSS condition means that you really don't need to worry about the three corresponding pairs of angles. Knowing that the three corresponding pairs of sides are congruent is sufficient for the triangles to be congruent. The three corresponding pairs of angles will necessarily match up if the three corresponding pairs of sides are congruent.

* Similarly, SAS (side-angle-side) means that if two sides of a triangle, and the included angle between those two sides, are respectively congruent to two sides and the included angle of another triangle, then those two triangles are congruent. The other corresponding pair of sides and the other two corresponding pairs of angles will necessarily have to be congruent.

* Finally, ASA stands for angle-side-angle and means that if two angles of one triangle and the included side between them are respectively congruent to two angles and the included side of another triangle, then the two triangles are congruent. The other two corresponding pairs of sides and the other corresponding pair of angles will also have to be congruent.

– How do we know that these are sufficient criteria to determine triangle congruence?
  The way I like to think of it is that ASA, SAS, and SSS are sufficient criteria to determine a triangle "uniquely."

  * First, let's consider ASA. If you were told two angles and an included side, how many "different" triangles could you make? Well, in any of those triangles, you could orient the included side to be horizontal (and call it the base), which might involve a rotation (an isometry). You could reflect across the base (another isometry) if necessary so that the known angles are above the horizontal side. Finally, we could reflect in the perpendicular bisector of the base, if necessary, (another isometry) so that the larger of the two angles is at the left endpoint of the side. (If the two angle measures are equal, then no reflection is necessary.) Then our triangle must look something like this.

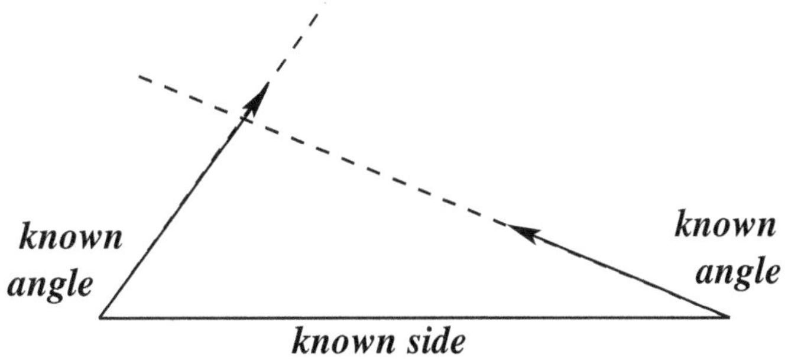

Now we really don't have any information about the other two sides, or the angle included between them, but we do know the angle that each of those sides makes with the base. So if we draw the two lines on which those sides lie, then they must intersect, and their intersection is at a point. That point must be the third vertex of the triangle, and so there is really only one triangle that has the ASA information we started with. Therefore, if two triangles shared ASA data, then we could perform rigid motions so that one triangle coincides with the other. Hence the triangles would be congruent.

* Second, we'll look at SAS. Suppose you know two side lengths and the measure of the included angle. Rotate the triangle so that the larger of the two sides is horizontal, the included angle is at the left, and the smaller of the two sides is above the horizontal side. (If the two sides have the same length, then it doesn't matter which one is horizontal.) A reflection might be needed to accomplish this.

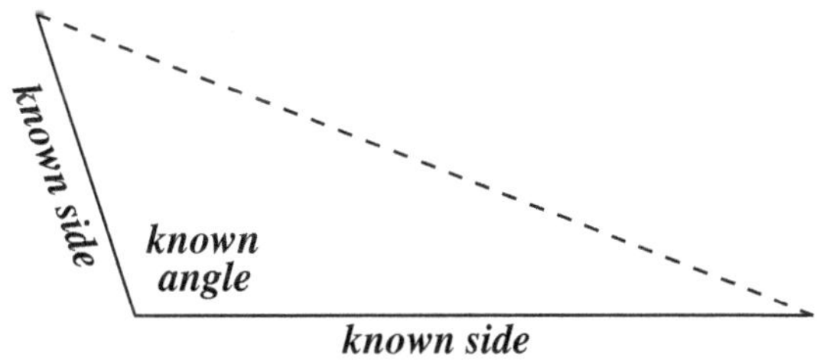

We do not know the other side length, but we do know where its two endpoints are, namely, the two free ends of the two sides we started with. The only way to complete this triangle is to join those two endpoints. Thus the third side and the two angles adjacent to the third side are completely determined by SAS. Therefore, if two triangles shared SAS data, then we could perform rigid motions so that one triangle coincides with the other. Hence the triangles would be congruent.

* Finally, we consider SSS. As before, let's rotate the triangle so that the largest

side is horizontal, and the next largest side meets the largest side at the left endpoint and lies above the largest side. You may need rotations and a reflection to accomplish this. (If all three sides are equal, then select one side for the base and another to meet at the left endpoint of the base and lie above the base. If two sides are equal and larger than the third side, then place either of the larger sides as the base and let the other large side meet it at its left endpoint. If two sides are equal and smaller than the third side, then still place the largest side as the base and use either of the other two as the "left" side.) We'll show two different cases below.

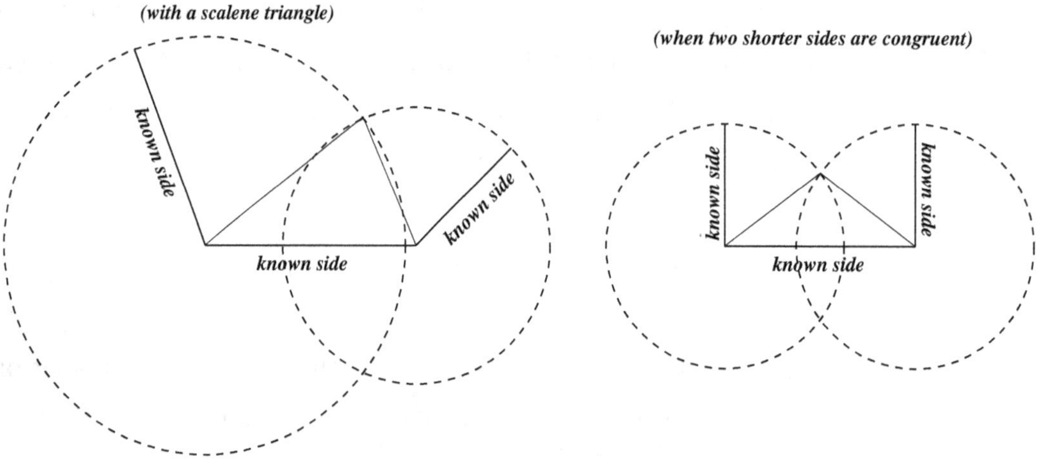

We do not know any of the angles. So we can think of the two "shorter" sides as being attached to the largest side on hinges. They can rotate in a circle around their fixed endpoints. To make a triangle, these two sides must meet at the third vertex of the triangle. But there is only one point above the horizontal where these two circles intersect. Hence there is only one triangle with given SSS data. That means if someone else made another triangle with those lengths, then your two triangles would be congruent, because you could transform one to the other through rigid motions.

Incidentally, what if the two shorter sides don't reach each other, like if you had side lengths of 3, 4, and 8?

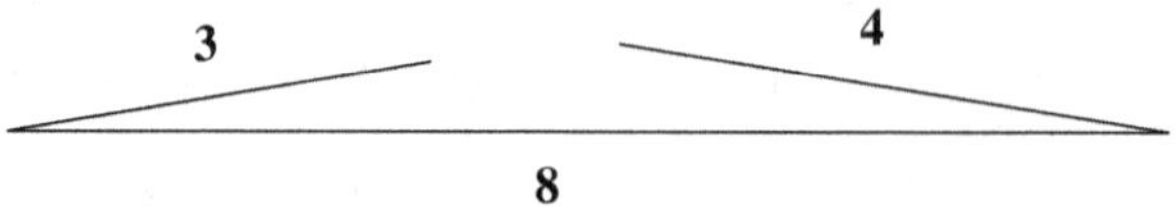

Then we have inadvertently hit upon the Triangle Inequality, which states that in any triangle, the sum of two side lengths must be longer than the third side. Stated another way, it says that if the sum of two side lengths is less than or equal to the third side, then you cannot form a triangle with those three side lengths.

– Are the other criteria, like AAS, AAA, or SSA, sufficient to determine triangle congruence?
Again, the way I think about it is to ask in each case whether the given data are sufficient to uniquely determine a triangle. The first one here, AAS, is actually sufficient to determine triangle congruence, and that's because it's equivalent to ASA. Why is that? Well, because in any Euclidean triangle, the angle sum is 180 degrees. (Proof given in #10, below.) So, if you know two angles, then you really know all three angles, because the third angle measure equals 180 minus the sum of the other two angles. Then that means you really know the angles on either side of the given side. Thus if you know AAS, you also know ASA. Similarly, if you know ASA, you also know AAS because you can determine the other angle. Therefore, AAS and ASA are completely equivalent.
Perhaps with AAA and SSA, it's easiest just to show examples in each case of two triangles that share the given data, but are clearly not congruent.
With AAA, you can have two triangles of different sizes. These will be called "similar" triangles, as we will see later.

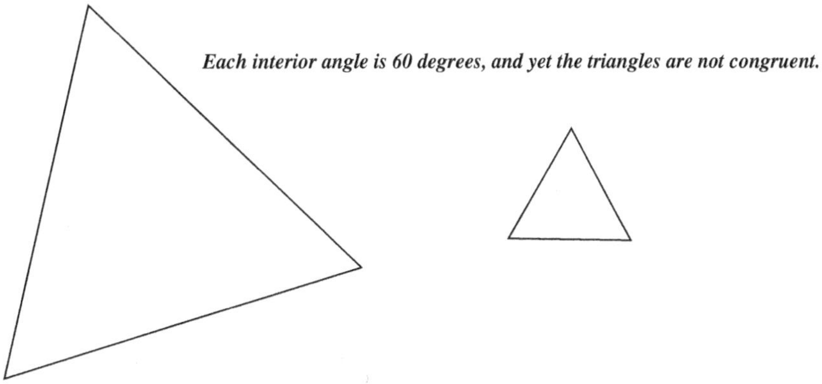

Each interior angle is 60 degrees, and yet the triangles are not congruent.

With SSA (or its snicker-worthy equivalent ASS), we have the possibility for what is sometimes called "the ambiguous case." It's possible for two triangles to share two side lengths and the angle at the endpoint of one of those sides, and yet be different triangles. In the picture below, $\angle C \cong \angle C'$ and $BC = BC'$.

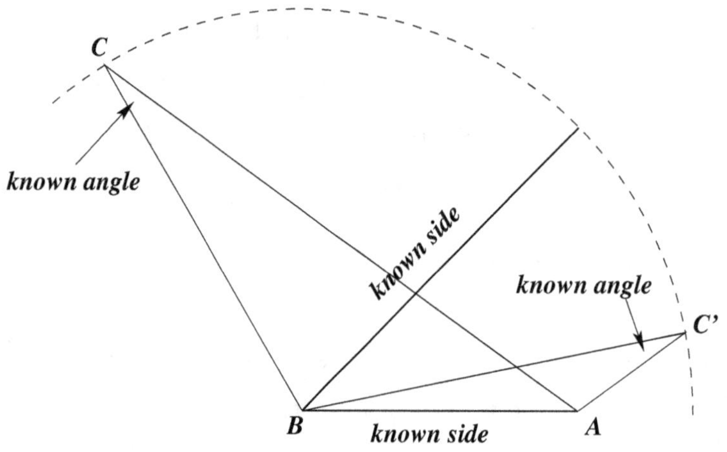

Triangles ABC and ABC' share SSA data, yet are not congruent.

Therefore, AAA and SSA are not sufficient to determine triangle congruence.

– Are there any more ways to show triangles are congruent?

There is a way to show that two right triangles are congruent, called Hypotenuse-Leg, abbreviated HL. This one can be proved, once we have proved the Pythagorean Theorem, which says that the square of the hypotenuse equals the sum of the squares of the two legs. So if you know the hypotenuse and one of the legs, then there is only one unknown variable in the Pythagorean Theorem, which would allow you to uniquely determine the other leg. Since there is only one triangle with given HL data, then any two triangles that share HL data must be congruent. There are other ways to prove HL, but we use it so seldom in what follows that we won't discuss other proofs here. Consult a good Geometry text, or look online, for more details.

- Prove geometric theorems

  – Why do we prove geometric theorems?

  The idea of "proof" in geometry is many things. Geometrical proof is a way to justify that your answer is correct. In this regard, proof is akin to checking your answer in other math courses. Proving an answer also involves higher-level thinking skills than just performing steps. In terms of Bloom, for instance, proving a statement requires evaluative judgements based on reasoning, which is different than just performing "plug and chug" calculations.

A more constructivist view of proof might be as the method of building mathematical knowledge upon itself. By starting from humble beginnings, such as undefined notions of points and lines, we can build up the structure of geometrical theorems. Proof is the mortar that holds the structure together. By proving a statement, that statement becomes a theorem which can then be used to build even more theorems. Without proof, we would not be able to justify whether statements are true or not.

Ideally, proof is used more and more in higher mathematics courses as students progress through their learning careers. There even comes a point in mathematics when the students realize that they themselves can determine if an answer is correct. Once they gain the critical thinking skills needed to evaluate a mathematical proof, then they will be able to critique their own work, and thus have confidence that they have deeply understood the material, and not merely mastered its manipulation.

Proving geometry theorems is an important step on the road toward conceptual understanding in mathematics.

– How do we prove geometric theorems?

We have discussed earlier that many theorems in mathematics can be phrased as conditional, or "if-then" statements. The way to prove such a statement is to assume that the hypotheses are true, and then try to deduce the conclusion from these true statements, as well as from undefined terms, definitions, axioms, postulates, and previously established theorems.

– What assumptions can me make? What earlier material can we use?

We have already established a definition of congruence that relies on rigid motions (isometries). When needed, we may also use various algebraic properties of real numbers in proofs. Also, we will use two key definitions: a midpoint of a line segment divides the segment into two congruent segments; similarly, an angle bisector divides an angle into two congruent angles.

Since this course is in Euclidean geometry, we can also use the Parallel Postulate, often stated as, "given a line and a point not on that line, there is exactly one line through the given point that is parallel to the given line." We will treat this as an axiom. That means we can accept it as true, without proof.

9. Prove theorems about lines and angles. *Theorems include: vertical angles are congruent; when a transversal crosses parallel lines, alternate interior angles are congruent*

and corresponding angles are congruent; points on a perpendicular bisector of a line segment are exactly those equidistant from the segment's endpoints.

– What are "vertical angles"? How do you prove vertical angles are congruent?

  Vertical angles are formed when two lines intersect. Vertical angles share a common vertex but not common sides, although their sides taken together do make up the two lines of intersection. Vertical angles lie opposite each other when two lines intersect. In the picture below, angles 1 and 3 are vertical, as are angles 2 and 4.

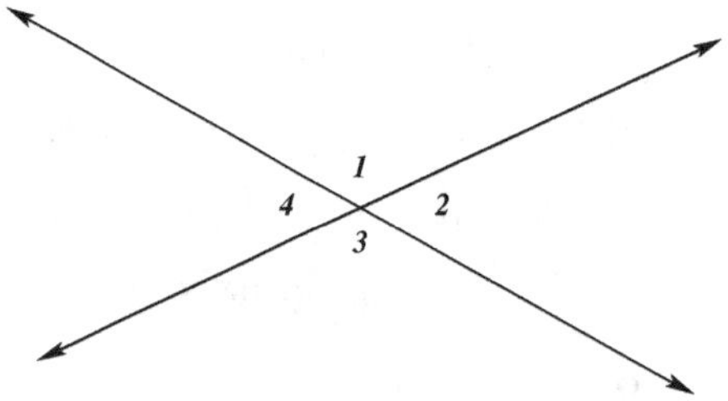

  Vertical angles are indeed congruent. Using our definition of congruence given above, we see that if we rotate angle 1 180 degrees around its vertex, then it exactly lines up on top of angle 3. Since rotation is a rigid motion (an isometry), the two figures must be congruent. The same thing is true of angles 2 and 4.

– What is a "transversal"?

  A transversal is a line that intersects each of two given lines. The two points of intersection are distinct. In the picture below, line t is a transversal, intersecting lines k and m. Notice that eight angles are formed.

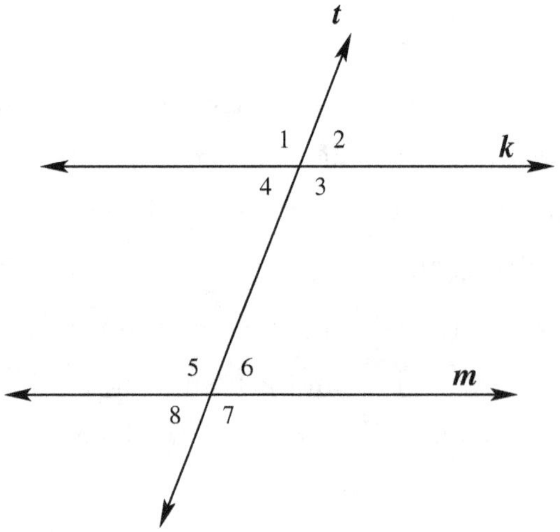

– What else do we need to know before we continue?

  Well, we really need two more facts in order to do justice to the logical nature of geometry, building only on what has come before. First, we will show that an exterior angle of a triangle is bigger than either remote interior angle. Second, we will show that if two lines are cut by a transversal in such a way that alternate interior angles are congruent, then the lines must be parallel.

FIRST FACT: an exterior angle of a triangle is necessarily bigger than either remote interior angle. By "bigger" we will show that one angle is entirely contained inside the other. We will prove it here, following Euclid's own proof. Interestingly, this is even true in so-called "non-Euclidean" geometries.

Consider $\triangle ABC$, and extend $\overline{AB}$ beyond point B. Label a point D on this extension, so that $\angle DBC$ is an exterior angle of $\triangle ABC$. We will show that $\angle DBC > \angle ACB$.

Identify point E as the midpoint of $\overline{BC}$. Thus $\overline{CE} \cong \overline{BE}$. Draw $\overleftrightarrow{AE}$ and locate point F on $\overleftrightarrow{AE}$ so that E is the midpoint of $\overline{AF}$. Thus $\overline{AE} \cong \overline{FE}$. We also know that $\angle AEC \cong \angle FEB$ because they are vertical angles.

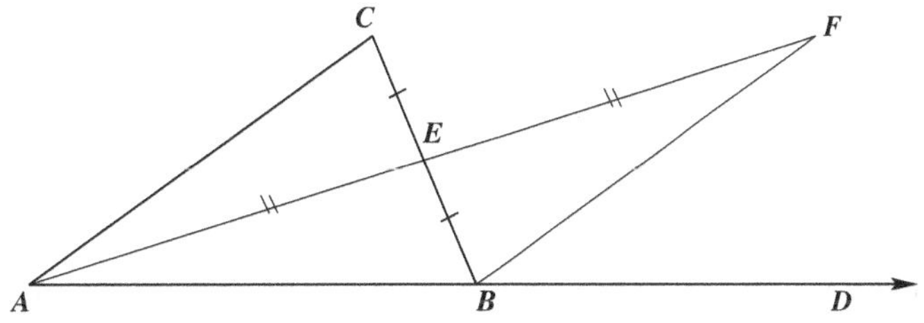

Therefore, by SAS, $\triangle AEC \cong \triangle FEB$, and so it follows that $\angle FBE \cong \angle ACE$. But $\angle ACE = \angle ACB$, and it's clear that $\angle FBE < \angle DBC$. So we have

$$\angle ACB = \angle ACE \cong \angle FBE < \angle DBC,$$

which was what we wanted to show. A similar drawing would show that the exterior angle at B is also bigger than $\angle CAB$. This ends the proof of the first fact.

SECOND FACT: If two lines are cut by a transversal in such a way that alternate interior angles are congruent, then the lines must be parallel. Recall that "parallel" in planar geometry just means that the lines do not intersect.

To show this, we will use an indirect proof method: proof by contradiction. That is, we will assume that the hypotheses are true, AND that the lines do in fact intersect. Then we will show that this situation is not possible. Suppose that $\overleftrightarrow{AB}$ and $\overleftrightarrow{A'B'}$ are cut by transversal $\overleftrightarrow{AA'}$ in such a way that $\angle BAA' \cong \angle AA'D$ AND that $\overleftrightarrow{AB}$ and $\overleftrightarrow{A'B'}$ intersect at point C. (If the point of intersection had been on the other side of the transversal, then we could just rotate the picture 180 degrees until it looks like the one below.)

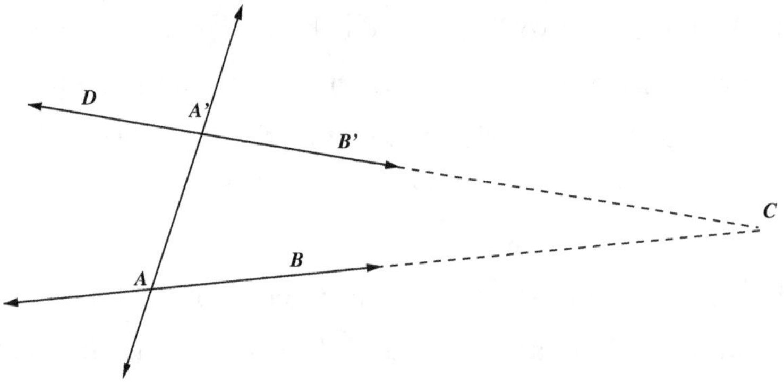

But notice that we would then have a triangle, $\triangle ACA'$, in which the exterior angle, $\angle DA'A$, is congruent to a remote interior angle, $\angle CAA'$. This is impossible, because we just got through proving that an exterior angle is bigger than either remote interior angle.

This contradiction means that $\overleftrightarrow{AB}$ and $\overleftrightarrow{A'B'}$ cannot intersect. Therefore they are parallel. This ends the proof of the second fact.

– How do you prove when a transversal crosses parallel lines, alternate interior angles are congruent and corresponding angles are congruent?

Now we will assume that the two lines are parallel, and deduce that the alternate interior angles are congruent. Notice that this is essentially the converse of the second fact we proved, above.

Suppose that $\overleftrightarrow{AB}$ is parallel to $\overleftrightarrow{A'B'}$. Let M be the midpoint of $\overline{AA'}$. Pick point C on $\overleftrightarrow{AB}$ and draw $\overleftrightarrow{CM}$. We know that $\overleftrightarrow{CM}$ must intersect $\overleftrightarrow{A'B'}$ at some point. Let's call that point D.

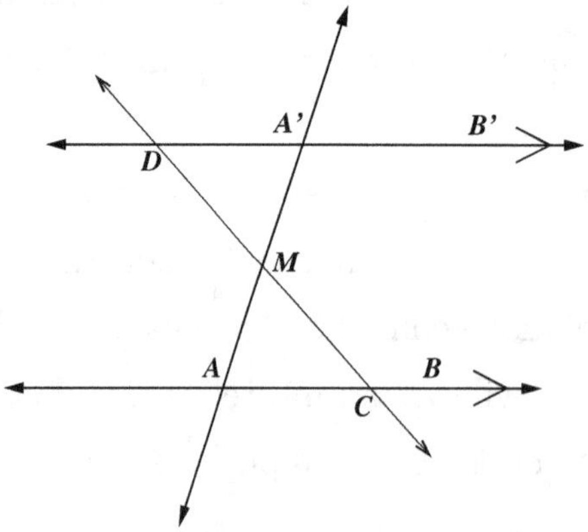

Now wait a minute, you might say. How do we know for sure that $\overleftrightarrow{CM}$ will intersect $\overleftrightarrow{A'B'}$ at all? That's a GREAT question, because you really shouldn't rely on my picture - you should rely on the logical reasoning. So let's temporarily suppose that $\overleftrightarrow{CM}$ does NOT intersect $\overleftrightarrow{A'B'}$. In geometry, that means $\overleftrightarrow{CM}$ is parallel to $\overleftrightarrow{A'B'}$. But then we would be contradicting the Parallel Postulate. According to the Parallel Postulate, there can only be one line through point C that is parallel to $\overleftrightarrow{A'B'}$. We would have two such lines: $\overleftrightarrow{AB}$ (given as parallel in the hypotheses) and $\overleftrightarrow{CM}$ (which is clearly different than $\overleftrightarrow{AB}$ because it contains point M). So, $\overleftrightarrow{CM}$ cannot be parallel to $\overleftrightarrow{A'B'}$. That's why we know that $\overleftrightarrow{CM}$ and $\overleftrightarrow{A'B'}$ must intersect at a point, which we called D.

OK, so now the picture has been justified. Let us now perform a rigid motion, namely a rotation of 180 degrees around point M. So, this rotation takes point A to A' because $\overline{AM} \cong \overline{A'M}$ and because M is on line $\overleftrightarrow{AA'}$. Rotating a line 180 degrees around a point on that line brings you back to the same line. (After all, 180 degrees is the measure of a straight angle.)

Where does C wind up after this rotation around M? In other words, where is C', the image of C under this rigid motion? Might C' be located as in the picture below?

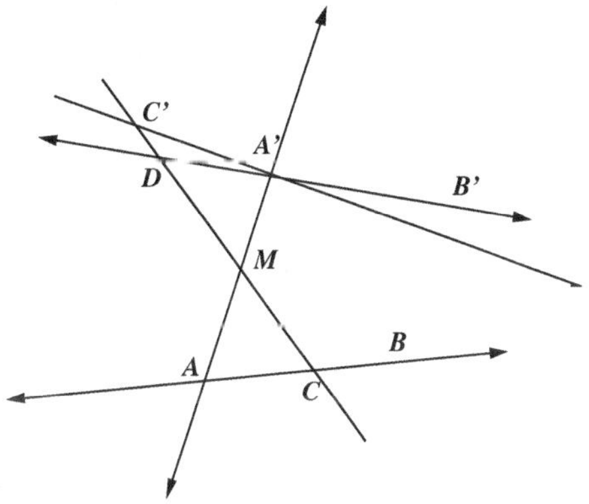

We will show that this picture cannot happen. First, we know that $\angle MAC \cong \angle MA'C'$ because rigid motions preserve angles. But these two angles are alternate interior angles to the two lines $\overleftrightarrow{AC}$ and $\overleftrightarrow{A'C'}$ with transversal $\overleftrightarrow{AA'}$. By our second fact above, then, we know that $\overleftrightarrow{A'C'}$ is parallel to $\overleftrightarrow{AC}$. But $\overleftrightarrow{A'B'}$ is the unique line through A' that is parallel to $\overleftrightarrow{AC}$. So this means that C' must lie on $\overleftrightarrow{A'B'}$. Second, we know that because the rotation is 180 degrees, $\overleftrightarrow{CM}$ will rotate to $\overleftrightarrow{CM}$ again. So C' must lie on $\overleftrightarrow{CM}$.

So we have shown that C' must lie on $\overleftrightarrow{A'B'}$ and on $\overleftrightarrow{CM}$, two lines which intersect at D. Since two lines can only intersect in one point, we have just proved that $C' = D$. So then it is true that $\angle MAC \cong \angle MA'C' = \angle MA'D$. Therefore, when two parallel lines are cut by a transversal, alternate interior angles are congruent. Fortunately, once we know that alternate interior angles are congruent, it is not too hard to see that corresponding angles are also congruent. This is because a corresponding angle can be described as a vertical angle of an alternate interior angle.

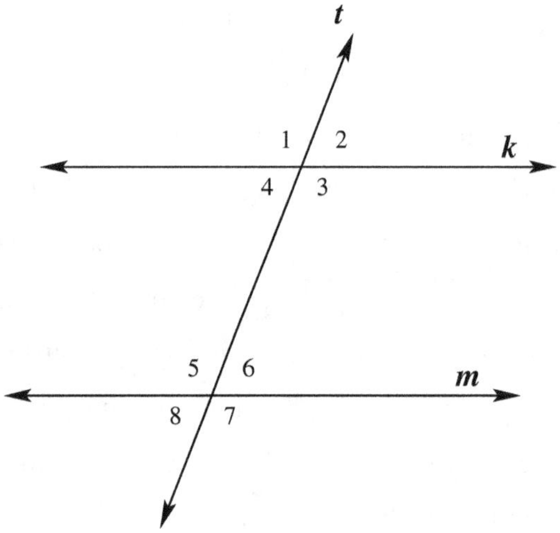

In this picture, we will show $\angle 1 \cong \angle 5$ using a conventional two-column format. We are given that line k is parallel to line m.

Statements	Reasons
$\angle 1 \cong \angle 3$	vertical angles are congruent
$\angle 3 \cong \angle 5$	alternate interior angles are congruent
$\angle 1 \cong \angle 5$	congruence is transitive

Other pairs of corresponding angles can similarly be shown to be congruent.

– Why didn't we just use the measures of the angles to come up with a proof? You could certainly do that. However, since we really haven't talked too much about measures of angles, I was trying to describe a proof that uses transformations, in the spirit of the Common Core State Standards.

– How do you prove points on a perpendicular bisector of a line segment are exactly those equidistant from the segment's endpoints?

  * What do they mean by "exactly"?
    Here, the word "exactly" is similar to a biconditional statement. It means that if a point is on the perpendicular bisector of a segment, then the point is equidistant from the endpoints of the segment. It also means that if a point is equidistant from the endpoints of a segment, then that point lies on the

perpendicular bisector of that segment. We must prove both statements.

∗ How do we prove each statement?

First, we will show that if a point lies on the perpendicular bisector of a line segment, then it is equidistant from the endpoints of the segment.

Consider $\overline{AB}$ with midpoint M and suppose that P lies on the perpendicular bisector of $\overline{AB}$. That means that $\overline{PM}$ is perpendicular to $\overline{AB}$. Now draw $\overline{PA}$ and $\overline{PB}$, making $\triangle PAM$ and $\triangle PBM$.

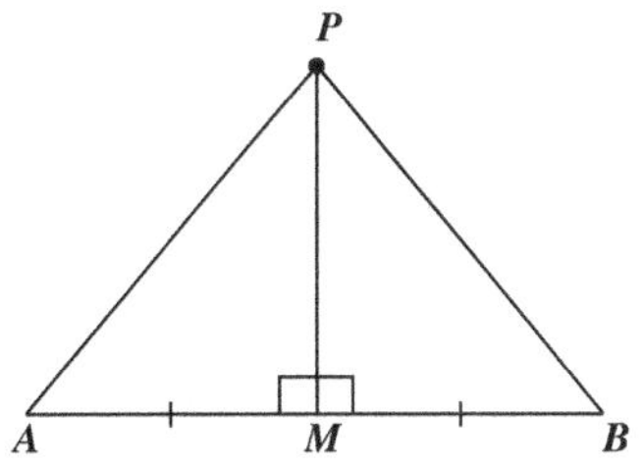

We know that $\overline{AM} \cong \overline{BM}$ by the definition of the midpoint. We also know that $\angle PMA \cong \angle PMB$ because they are both right angles and all right angles are congruent to each other. And certainly $\overline{PM} \cong \overline{PM}$ because every object is congruent to itself (because the identity transformation is an isometry). Therefore, by SAS, $\triangle PAM \cong \triangle PBM$. From here it follows that $\overline{PA} \cong \overline{PB}$, which means that P is equidistant from A and B.

And now, we will show the converse statement: that if a point is equidistant from the endpoints of a line segment, then it lies on the perpendicular bisector of the segment. Here we have a similar picture, except that we start with $\overline{AB}$ and its midpoint M, and we draw P so that $\overline{PA} \cong \overline{PB}$. Now we must show that $\overline{PM}$ is perpendicular to $\overline{AB}$.

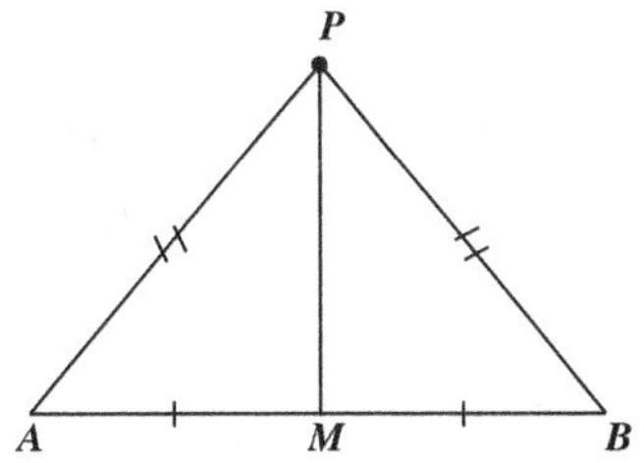

We know that $\overline{AM} \cong \overline{BM}$ by the definition of the midpoint. We also know that $\overline{PA} \cong \overline{PB}$ and that $\overline{PM} \cong \overline{PM}$. So, by SSS, $\triangle PAM \cong \triangle PBM$. Hence $\angle PMA \cong \angle PMB$. But we also know that $\angle PMA$ and $\angle PMB$ are supplementary. Congruent supplementary angles are right angles by definition. So $\overline{PM}$ is perpendicular to $\overline{AB}$, which means that P lies on the perpendicular bisector of $\overline{AB}$.

This concludes the proof that points on a perpendicular bisector of a line segment are exactly those points that are equidistant from the segment's endpoints.

10. **Prove theorems about triangles.** *Theorems include: measures of interior angles of a triangle sum to 180°; base angles of isosceles triangles are congruent; the segment joining midpoints of two sides of a triangle is parallel to the third side and half the length; the medians of a triangle meet at a point.*

   – How do you prove measures of interior angles of a triangle sum to 180°?

   There are a number of proofs of this. One of the most constructive is no actually create a triangle out of paper, cut out the three vertices and line them up to show that they comprise a straight line. Our proof will be an analytic version of this demonstration with manipulatives.

   Consider $\triangle ABC$. From the Parallel Postulate, we know that there is exactly one line through A that is parallel to $\overline{BC}$. Draw this line and place points D and E so that A is between D and E as in the picture.

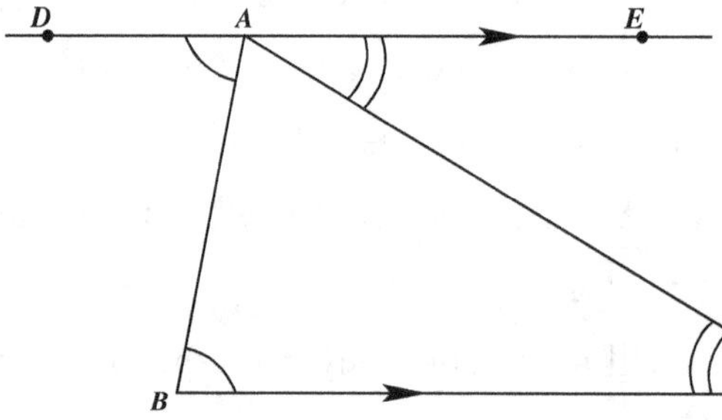

   Whenever two parallel lines are cut by a transversal, alternate interior angles are congruent. That means that $\angle DAB \cong \angle ABC$ and $\angle EAC \cong \angle ACB$. Looking at A, we see three angles that make up a straight line, or 180 degrees: $\angle DAB$, $\angle BAC$, and $\angle EAC$. Substituting congruent angles for two of these, we see that $\angle ABC$, $\angle BAC$, and $\angle ACB$ also make up a straight line. Therefore, the three interior angles of a triangle sum to 180°.

   – What does "isosceles" mean?

   The word "isosceles" comes from Greek words meaning "equal" and "leg." So an isosceles triangle has two sides that are congruent. If all three sides are congruent, the triangle is called "equilateral" (from Latin words for "equal" and "sides").

   – How do you prove base angles of isosceles triangles are congruent?

   Notice that the definition involves the side lengths of the triangle. To show that the base angles of an isosceles triangle are congruent, then, requires proof. It does not follow immediately from the definition. Also, notice that our assumption

below (that two sides are congruent) also applies to equilateral triangles. Hence, our conclusions will apply to equilateral triangles as well.
We will give two proofs here, because one of them uses symmetry in a more advanced way than the other. In each case, we are given $\triangle ABC$ with $\overline{AB} \cong \overline{AC}$, which makes $\overline{BC}$ the base. We must show that $\angle B \cong \angle C$.

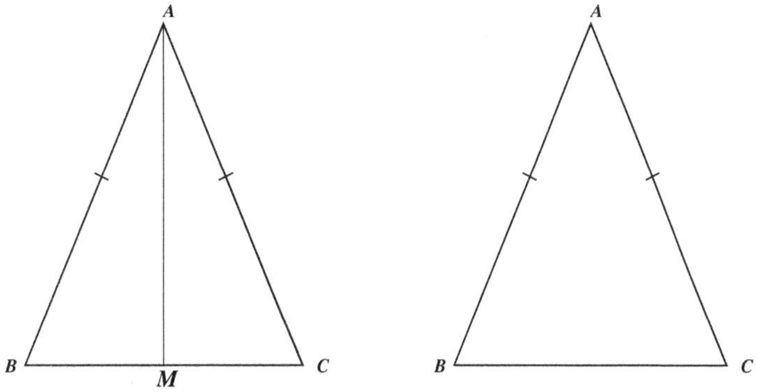

First Proof ***Second Proof***

* First Proof
  Since $\overline{AB} \cong \overline{AC}$, we know that A lies on the perpendicular bisector of $\overline{BC}$. Let M be the midpoint of $\overline{BC}$ and draw perpendicular bisector $\overline{AM}$. So we have shown, by SSS, that $\triangle ABM \cong \triangle ACM$. (Remember what the definition of "midpoint" is.) Therefore, $\angle B \cong \angle C$, because corresponding parts of congruent triangles are congruent.
* Second Proof
  We know that $\overline{AB} \cong \overline{AC}$ and clearly $\angle A \cong \angle A$. Therefore, by SAS, $\triangle BAC \cong \triangle CAB$. Therefore, $\angle B \cong \angle C$, because corresponding parts of congruent triangles are congruent.
  This second proof is rather subtle. It uses the natural symmetry of an isosceles triangle, and it highlights the fact that when you say that two triangles are congruent, the order in which the vertices are listed is very important. We have just seen that a triangle can be congruent to itself, but via a transformation that is not just the identity. Here, we could explain this proof by saying that we have essentially reflected the triangle in the perpendicular bisector of $\overline{BC}$, thus exhibiting a symmetry of the original triangle. Since $\angle B$ is coincident with $\angle C$ after that reflection, the angles must be congruent.

- How do you prove the segment joining midpoints of two sides of a triangle is parallel to the third side and half the length?

There are many proofs, but I will try to use one that only relies on what we have used so far, mainly on triangle congruences.
Suppose that in $\triangle ABC$, M is the midpoint of $\overline{AB}$. Draw the unique line through M that is parallel to $\overline{BC}$. It must intersect $\overline{AC}$ at a point; call it N. (We will show later why N has to be the midpoint of $\overline{AC}$, but we have not proven it yet.) Similarly, draw the unique line through M that is parallel to $\overline{AC}$. It intersects $\overline{BC}$ at point P. (Again, we have not yet established that P is the midpoint of $\overline{BC}$.) Also, draw $\overline{PN}$.

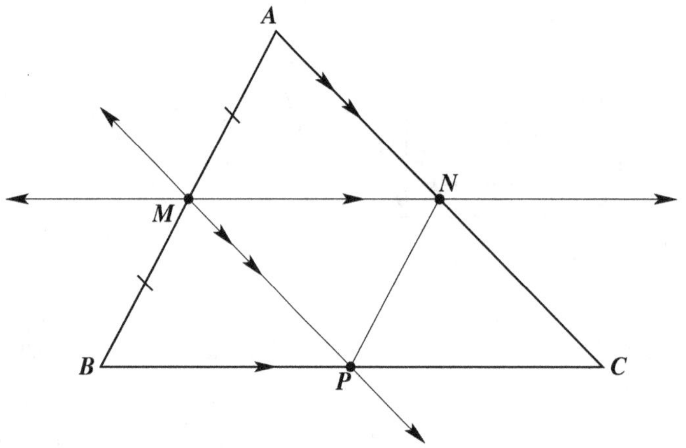

We will proceed in steps:

(a) $\triangle AMN \cong \triangle MBP$ by ASA. ($\angle A \cong \angle BMP$ and $\angle AMN \cong \angle B$ because when parallel lines are cut by a transversal, corresponding angles are congruent; $\overline{AM} \cong \overline{MB}$ by definition of midpoint)

(b) $\overline{AN} \cong \overline{MP}$ and $\overline{MN} \cong \overline{BP}$ because corresponding parts of congruent triangles are congruent.

(c) $\triangle CNP \cong \triangle MPN$ by ASA. ($\angle CPN \cong \angle MNP$ and $\angle CNP \cong \angle MPN$ because when parallel lines are cut by a transversal, alternate interior angles are congruent; $\overline{PN} \cong \overline{PN}$)

(d) $\overline{MN} \cong \overline{CP}$ and $\overline{MP} \cong \overline{CN}$ because corresponding parts of congruent triangles are congruent.

(e) $\overline{AN} \cong \overline{CN}$ and $\overline{BP} \cong \overline{CP}$ by the transitive property of congruence.

(f) N is the midpoint of $\overline{AC}$ and P is the midpoint of $\overline{BC}$ by the definition of midpoint.

(g) Therefore, the line through M that is parallel to $\overline{BC}$ is exactly the line that joins M to the midpoint of $\overline{AC}$.

(h) Moreover, $\overline{BC}$ is divided by P into two segments, each of which is congruent to $\overline{MN}$. Therefore the length of $\overline{MN}$ is one half the length of $\overline{BC}$.

We are done.

- What is a "median" of a triangle?

  A median of a triangle is a line segment joining a vertex to the midpoint of the opposite side. Hence there are three distinct medians in any triangle.

- How do you prove the medians of a triangle meet at a point?

  We need to show that there is a point that lies on all three lines. (By the way, this point is called the "centroid" of the triangle.) A quick internet search shows many different proofs, but again, we will try to find one that relies only on what we have already shown. The following proof can be found at http://www.cut-the-knot.org/triangle/medians.shtml.

  But first, we will prove a slightly different statement, one with more specificity in a way. This specificity will turn out to play a key role in the proof.

  We will prove that any two medians of a triangle intersect at a point (called the centroid) that is twice as far from each vertex as it is from the midpoint of that vertex's opposite side. In other words, the centroid lies at a distance from a vertex that is two-thirds the length of that vertex's median. By specifying exactly where the centroid lies on each median, we are guaranteeing that if any two medians intersect at that point, then all three must intersect at that point.

  Given: $\triangle ABC$ with medians $\overline{CP}$ and $\overline{BN}$, intersecting at Q.

  Prove: $PQ = \frac{1}{3}CP$ and $NQ = \frac{1}{3}BN$.

  To begin, let's draw $\overline{PN}$. Also, locate S as the midpoint of $\overline{BQ}$ and T as the midpoint of $\overline{CQ}$, and draw $\overline{ST}$.

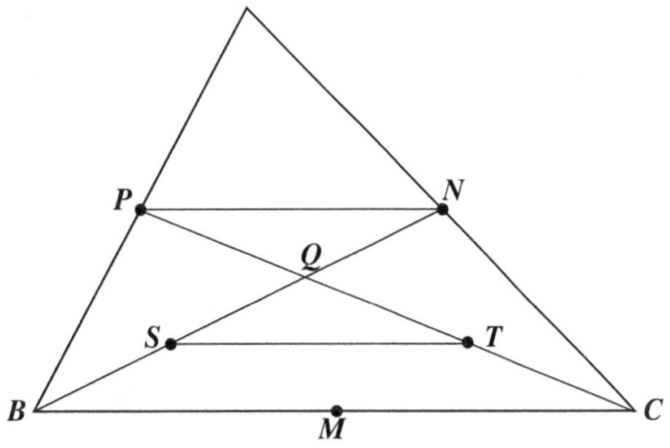

  Because P and N are midpoints, we know that $\overline{PN}$ is parallel to $\overline{BC}$ and $PN =$

$\frac{1}{2}BC$. Similarly in $\triangle BQC$, $\overline{ST}$ is parallel to $\overline{BC}$ and $ST = \frac{1}{2}BC$. So $\overline{ST}$ is also parallel to $\overline{PN}$ and $\overline{ST} \cong \overline{PN}$. Since alternating interior angles are congruent, we have $\angle TSQ \cong \angle PNQ$ and $\angle STQ \cong \angle NPQ$. So, by ASA, $\triangle STQ \cong \triangle NPQ$. Since corresponding parts of congruent triangles are congruent, we have $\overline{PQ} \cong \overline{TQ}$. Further, $\overline{TQ} \cong \overline{TC}$ because T is the midpoint of $\overline{QC}$. Thus $PQ = \frac{1}{3}CP$. In a similar way, $NQ = \frac{1}{3}BN$. This completes the proof.
Since these two medians were arbitrary, we know that Q also lies on the median through A and the distance from Q to A is also two-thirds the length of that median.

11. **Prove theorems about parallelograms.** *Theorems include: opposite sides are congruent, opposite angles are congruent, the diagonals of a parallelogram bisect each other, and conversely, rectangles are parallelograms with congruent diagonals.*

- What is the definition of a parallelogram?
  A parallelogram is a quadrilateral that has both pairs of opposite sides parallel. Sometimes people confuse the definition with the properties that are deduced from the definition, but here, the word "parallelogram" should really make you think of parallel lines.

- How do you prove opposite sides of a parallelogram are congruent? How do you prove opposite angles of a parallelogram are congruent?
  We will examine these first two together because they follow from a relatively simple application of congruent triangles.
  Given: parallelogram $ABCD$.
  Prove that $\angle B \cong \angle D$ and $\overline{AB} \cong \overline{CD}$. (Notice that since we haven't specified anything else, this essentially proves that in a parallelogram, any two opposite angles are congruent and any two opposite sides are congruent.) We will use the conventional two-column format here. Begin by drawing $\overline{AC}$.

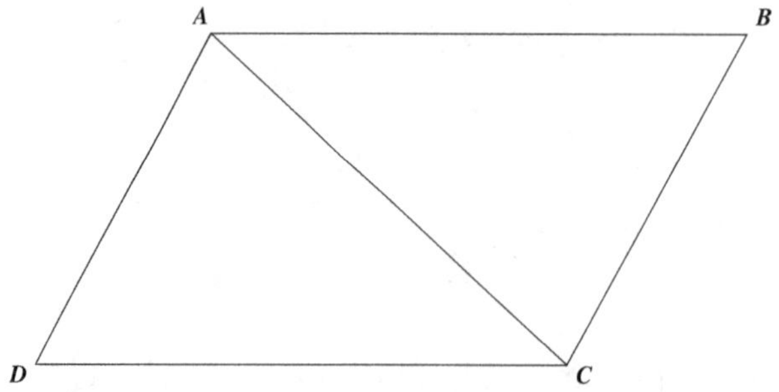

Statements	Reasons
$\overline{AB}$ is parallel to $\overline{CD}$; $\overline{AD}$ is parallel to $\overline{BC}$	definition of parallelogram
$\angle BCA \cong \angle DAC$; $\angle BAC \cong \angle DCA$	alt. int. angles are $\cong$
$\overline{AC} \cong \overline{AC}$	reflexive property of congruence
$\triangle ABC \cong \triangle CDA$	ASA
$\angle B \cong \angle D$; $\overline{AB} \cong \overline{CD}$	corr. parts of $\cong$ $\triangle$s are $\cong$

– How do you prove the diagonals of a parallelogram bisect each other, and conversely?

We have two statements to prove: (a) if $ABCD$ is a parallelogram, then $\overline{AC}$ and $\overline{BD}$ bisect each other; and (b) if $\overline{AC}$ and $\overline{BD}$ bisect each other in quadrilateral $ABCD$, then $ABCD$ is a parallelogram. Before we begin, suppose that the diagonals of $ABCD$ intersect at point E.

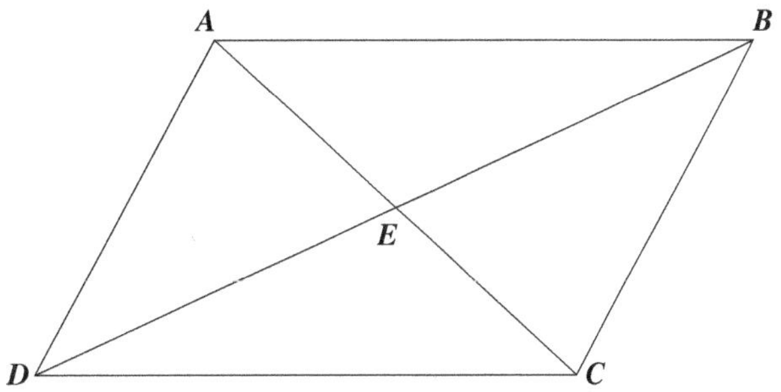

(a) Given: $ABCD$ is a parallelogram. Prove: $\overline{AC}$ and $\overline{BD}$ bisect each other.
From the definition of a parallelogram, we know that $\overline{AB}$ and $\overline{CD}$ are parallel. That means that pairs of alternate interior angles are congruent; thus, $\angle BAE \cong \angle DCE$ and $\angle ABE \cong \angle CDE$. We also know that $\overline{AB} \cong \overline{CD}$ because opposite sides of a parallelogram are congruent. So, by ASA, $\triangle ABE \cong \triangle CDE$. From this, we deduce that $\overline{AE} \cong \overline{CE}$ and $\overline{BE} \cong \overline{DE}$. Therefore, the diagonals of the parallelogram bisect each other.

(b) Given: $\overline{AC}$ and $\overline{BD}$ bisect each other. Prove: $ABCD$ is a parallelogram.
For the converse, we are given that $\overline{AC}$ and $\overline{BD}$ bisect each other. That means that $\overline{AE} \cong \overline{CE}$ and $\overline{BE} \cong \overline{DE}$. Also, at E, we have some vertical angles, which are thus congruent, namely, $\angle AEB \cong \angle CED$ and $\angle AED \cong \angle CEB$. So, using SAS in each case, we have that $\triangle AEB \cong \triangle CED$ and $\triangle AED \cong \triangle CEB$. Since congruent triangles have congruent corresponding parts, we know that $\angle ABE \cong \angle CDE$, which means that $\overline{AB}$ is parallel to $\overline{CD}$. Similarly, since $\angle ADE \cong \angle CBE$, $\overline{AD}$ is parallel to $\overline{CB}$. Therefore,

$ABCD$ is a parallelogram.

– How do you prove rectangles are parallelograms with congruent diagonals?

This statement is not phrased as an if-then statement. Using the word "are" should make you think that this is more like a biconditional. In other words, this sentence contains both a conditional statement – If $ABCD$ is a rectangle, then its diagonals are congruent. – and its converse – If parallelogram $ABCD$ has congruent diagonals, then it is a rectangle. We will prove both statements.

First, suppose that $ABCD$ is a rectangle. Then we know $ABCD$ is a parallelogram. (If in doubt, extend the sides and look at alternate interior angles.) So, from what has already been proven, $\overline{AB} \cong \overline{CD}$. Also, $\angle B \cong \angle C$ because both are right angles. Clearly $\overline{BC} \cong \overline{BC}$. Draw $\overline{AC}$ and $\overline{BD}$.

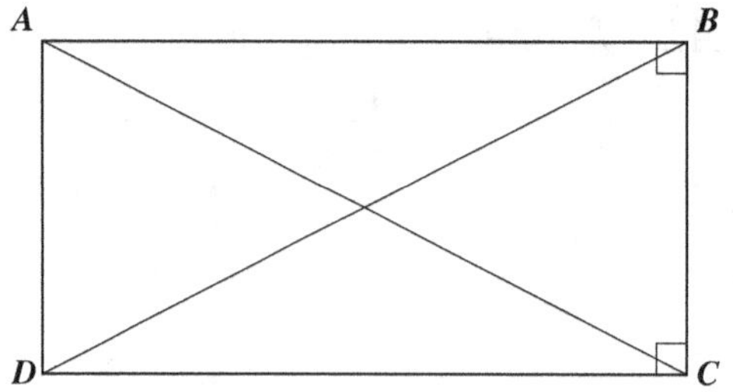

So, by SAS, $\triangle ABC \cong \triangle DCB$. And since corresponding parts of congruent triangles are congruent, $\overline{AC} \cong \overline{BD}$. So the diagonals of a rectangle are congruent.

Conversely, suppose that the diagonals of parallelogram $ABCD$ are congruent. That is, $\overline{AC} \cong \overline{BD}$. Because opposite sides of a parallelogram are congruent, $\overline{AB} \cong \overline{CD}$. Clearly, $\overline{BC} \cong \overline{BC}$. So, by SSS, $\triangle ABC \cong \triangle DCB$. Hence $\angle ABC \cong \angle DCB$. Extend $\overline{AB}$ beyond B to point E.

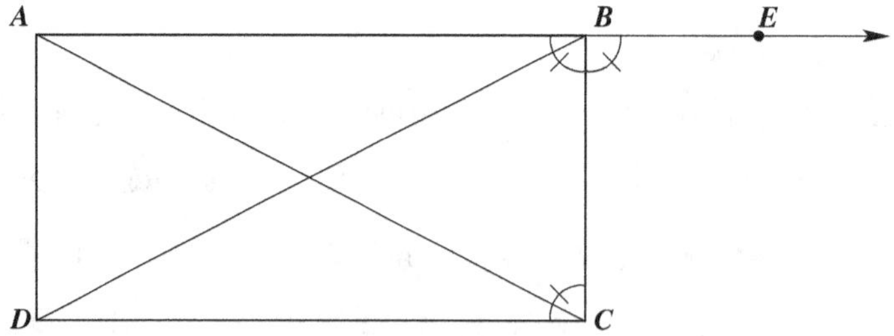

Since $\overline{AB}$ is parallel to $\overline{CD}$, we have $\angle DCB \cong \angle CBE$ because they are alternate interior angles. But then $\angle CBA \cong \angle CBE$ and $\angle CBA$ is supplementary to $\angle CBE$. This means that $\angle CBA$ and $\angle CBE$ are right angles. So $\angle DCB$ is also a right angle. Similar reasoning shows that $\angle DAB$ and $\angle ADC$ are also right angles. Therefore, a parallelogram with congruent diagonals is a rectangle.

This completes the proof.

- Make geometric constructions

12. Make formal geometric constructions with a variety of tools and methods (compass and straightedge, string, reflective devices, paper folding, dynamic geometric software, etc.). *Copying a segment; copying an angle; bisecting a segment; bisecting an angle; constructing perpendicular lines, including the perpendicular bisector of a line segment; and constructing a line parallel to a given line through a point not on the line.*

   – What makes a construction "classical?" What are some other approaches?

     The classical constructions are those that can be performed with a compass (for copying segments of a given fixed length) and an unmarked straightedge (for drawing or extending lines if you are given two points). These were the geometry tools used by the ancient Greeks.

     Other approaches involve string (which draws a circle, if attached at the center, and can draw an ellipse, if attached at the two foci), origami (the Japanese art of paper folding), reflective devices (such as a mirror), and of course software (such as the aforementioned GeoGebra or Geometer's Sketchpad). We will not go into how to use the software here.

     We will describe several classical constructions below, but will talk about other approaches when appropriate. If you think of attaching a piece of string at one point, then it becomes very similar to a compass, thus allowing you to do all the classical constructions with string and a straightedge.

   – How do you copy a segment? Why does it work?

     This is the building block of constructions. Using a compass, simply set the feet of the compass on the endpoints of $\overline{AB}$. Then, move the compass to A' and swing an arc on line $\overleftrightarrow{A'C'}$ to intersect it at B'. Now $\overline{A'B'} \cong \overline{AB}$. This works because the compass is assumed to be rigid, meaning that the distance between its feet do not change when it is moved from one point to another.

     Since folding paper involves a reflection (i.e., an isometry), then the image of a segment must be a congruent segment. So origami can be used to copy segments as well.

   – How do you copy an angle? Why does it work?

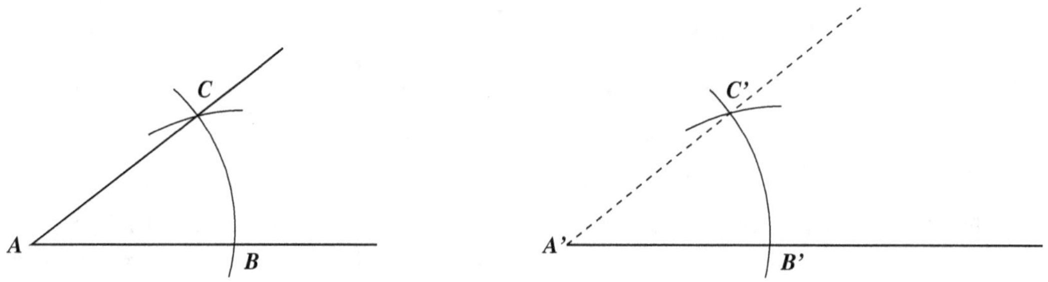

To copy the given $\angle A$ onto the given line at A', we first draw an arc through the sides of the angle. Label the points of intersection as B and C. Using the same compass setting, swing an arc centered at A'. Label its point of intersection with the line as B'. Now set the compass to the distance BC. Keeping the same compass setting, swing an arc centered at B' that intersects the first arc at a point C'. Then $\angle C'A'B' \cong \angle CAB$.

The reason that $\angle C'A'B' \cong \angle CAB$ is because $AB = AC = A'B' = A'C'$ and $BC = B'C'$. So by SSS, $\triangle C'A'B' \cong \triangle CAB$. Therefore, $\angle A \cong \angle A'$, because corresponding parts of congruent triangles are congruent.

As above, single folds of paper will also preserve angles. Angles can be copied via origami.

– How do you perpendicularly bisect a line segment (or construct a bisector of a segment, or construct perpendicular lines)? Why does it work?

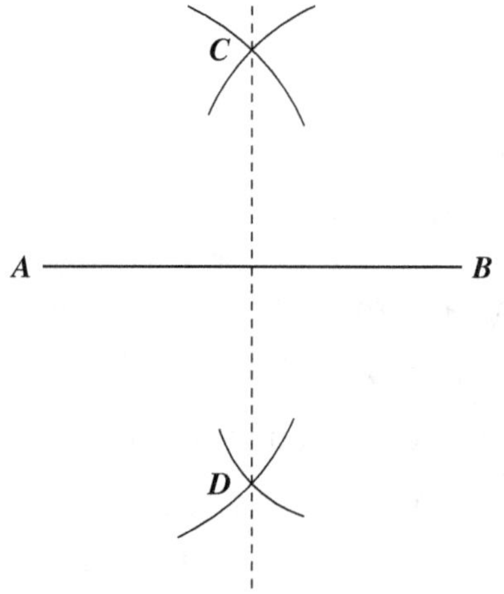

To perpendicularly bisect the given segment $\overline{AB}$, we first set the compass to a distance that is at least half of AB. Then swing arcs centered at A on both sides of $\overline{AB}$. Using the same compass setting, swing arcs centered at B on both sides of $\overline{AB}$ so that the new arcs intersect the old ones. Label the points of intersection C and D. Then $\overline{CD}$ perpendicularly bisects $\overline{AB}$.

The reason that $\overline{CD}$ perpendicularly bisects $\overline{AB}$ is that $AC = AD = BC = BD$, which makes $ADBC$ a rhombus. The diagonals of a rhombus perpendicularly bisect each other. This can also be shown using triangle congruences.
Using paper folding, you can simply fold the paper so that A lies on top of B. Then unfold. The crease will perpendicularly bisect $\overline{AB}$. There is also a way to create perpendicular lines. Since origami starts with a square piece of paper, folding the paper in half by bringing one side to its opposite side will create a crease down the middle that is perpendicular to each of the sides it intersects.

– How do you construct a line perpendicular to a given line through a point not on that line? Why does it work?
Though not explicitly mentioned on the Common Core State Standards, we will use this construction later on. We are given a line m and a point P not on m.

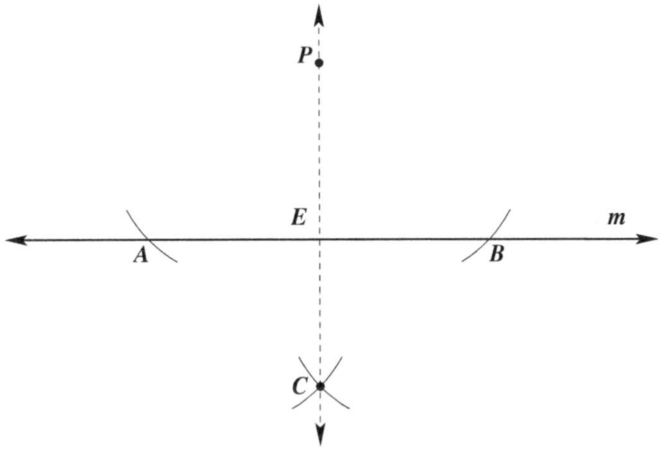

First, set the compass to a distance greater than the distance from P to the line and swing arcs intersecting m at A and B. Next, using a compass setting that is greater than half the distance between A and B, swing an arc at A and an arc at B that intersect at a point C other than P. It's probably easiest (though not necessary) to make this intersection happen on the side of m opposite P. Draw $\overleftrightarrow{PC}$ and let E denote its point of intersection with m. This line is perpendicular to m at E.
The reason that $\overleftrightarrow{PC}$ is perpendicular to m can be seen using congruent triangles. By construction, $PA = PB$ and $AC = BC$. From SSS, we know that $\triangle PAC \cong \triangle PBC$. That means that $\angle APE \cong \angle BPE$. So, by SAS, $\triangle PAE \cong \triangle PBE$, which means that $\angle PEA \cong \angle PEB$. Since these are congruent and supplementary, they must be right angles. Thus, $\overleftrightarrow{PC}$ is perpendicular

to m.

Using paper folding, the construction is slightly simpler. Simply fold the paper so that line m folds on top of itself, and point P is on the crease. Now unfold the paper. The crease will be perpendicular to m and will pass through point P, as desired.

– How do you bisect an angle? Why does it work?

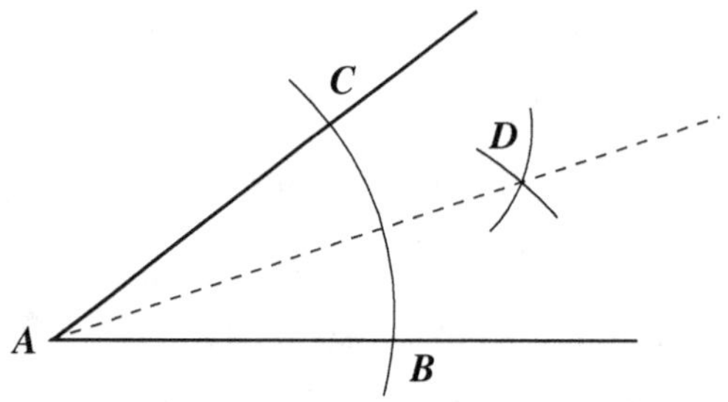

To bisect the given $\angle A$, we first draw an arc through the sides of the angle. Label the points of intersection as B and C. Using a new compass setting that is at least half the distance BC, swing two arcs, one centered at B and one centered at C, so that they intersect at a point D in the interior of angle A. Then $\overrightarrow{AD}$ bisects angle A.

The reason that $\overrightarrow{AD}$ bisects angle A is that $AB = AC$, $BD = CD$, and $AD = AD$. So, by SSS, $\triangle DAB \cong \triangle DAC$. Therefore, $\angle DAB \cong \angle DAC$, because corresponding parts of congruent triangles are congruent.

Using paper, simply fold the paper so that $\overrightarrow{AB}$ lies on top of $\overrightarrow{AC}$ (with A not moving – staying on the crease). Fold and then unfold. The crease will bisect $\angle BAC$.

– How do you construct a line parallel to a given line through a point not on that line? Why does it work?

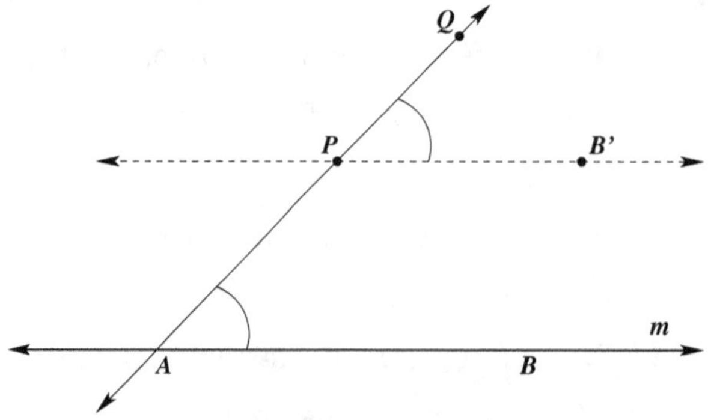

To construct a line through P that is parallel to $\overleftrightarrow{AB}$, first draw $\overleftrightarrow{AP}$ and on it, label point Q so that P is between A and Q. Now copy $\angle PAB$ on $\overrightarrow{PQ}$ so that $\angle QPB'$ (the copy) is a corresponding angle to $\angle PAB$. Then $\overleftrightarrow{PB'}$ is parallel to $\overleftrightarrow{AB}$.

The reason that $\overleftrightarrow{PB'}$ is parallel to $\overleftrightarrow{AB}$ is that we have constructed two lines cut by a transversal in such a way that corresponding angles are congruent. And if corresponding angles are congruent, then the lines are parallel.

– What is another way to construct a line parallel to a given line through a point not on that line? Why does it work?

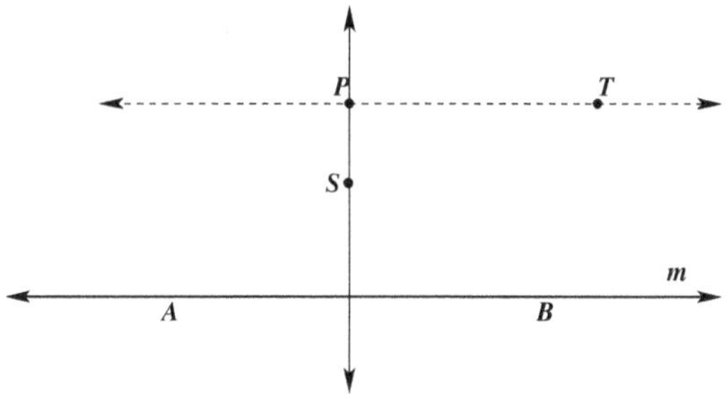

Using paper folding, we can take a different route. First, we fold the paper so that P is on the crease, and $\overleftrightarrow{AB}$ folds on top of itself. Then unfold. Now P is on a crease that is perpendicular to $\overleftrightarrow{AB}$. We have found the line through P that is perpendicular to $\overleftrightarrow{AB}$. Draw this line and label a point S on it. Now we fold the paper again, keeping P on the crease, but folding so that $\overleftrightarrow{PS}$ folds on top of itself. Unfold. The crease now gives a line through P that is perpendicular to $\overleftrightarrow{PS}$. Label another point T on this new line. Then $\overleftrightarrow{PT}$ is parallel to $\overleftrightarrow{AB}$.

The reason $\overleftrightarrow{PT}$ is parallel to $\overleftrightarrow{AB}$ is that we have constructed two lines that are crossed perpendicularly by a transversal. Since alternate interior angles are congruent (right) angles, the lines are parallel.

There are other valid constructions as well.

13. Construct an equilateral triangle, a square, and a regular hexagon inscribed in a circle.

– How do you construct a regular hexagon inscribed in a circle? Why does it work?

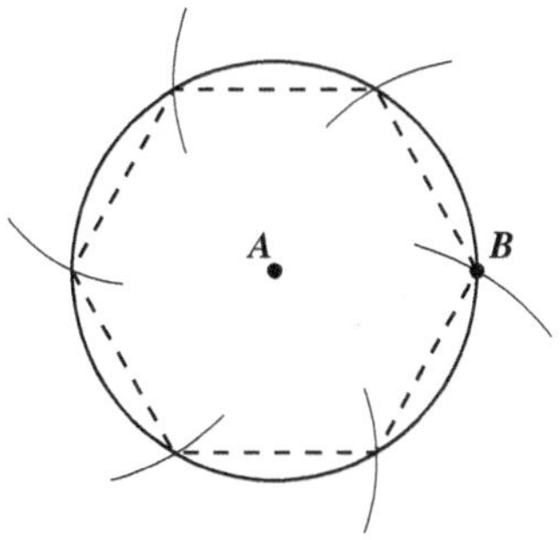

We are starting with the regular hexagon inscribed in a circle, because the equilateral triangle is easy to inscribe once you have the hexagon. To construct a regular hexagon, draw a circle centered at some point A. Select a point B on the circle. Using the same compass setting as you used to draw the circle, swing arcs around the circumference of circle A, starting at B. After six arcs, you should be back at B. The intersection points between the arcs and the circle are the vertices of a regular hexagon.
The reason this works is that, for a regular hexagon inscribed in a circle, the radius of the circle and the side length of the hexagon are equal.

– How do you inscribe an equilateral triangle in a circle?
Use the construction given above for a regular hexagon inscribed in a circle. Then connect every other vertex to form an equilateral triangle.

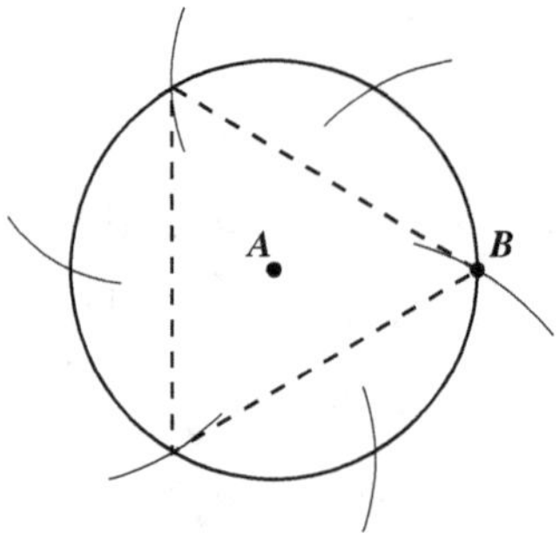

– How do you inscribe a square in a circle?
There are a number of ways to do this. One way is to draw a diameter of the circle, then perpendicularly bisect the diameter to find the perpendicular diameter. The four endpoints of these diameters are the vertices of a square. See the picture below.

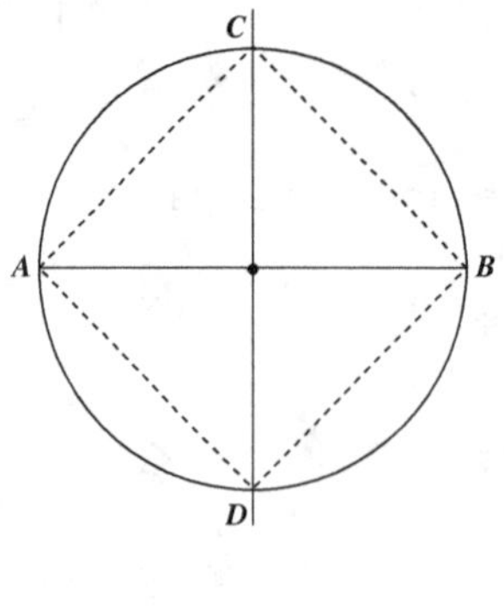

SIMILARITY, RIGHT TRIANGLES, AND TRIGONOMETRY

- UNDERSTAND SIMILARITY IN TERMS OF SIMILARITY TRANSFORMATIONS

1. Verify experimentally the properties of dilations given by a center and a scale factor:
   (a) A dilation takes a line not passing through the center of the dilation to a parallel line, and leaves a line passing through the center unchanged.
   (b) The dilation of a line segment is longer or shorter in the ratio given by the scale factor.

   – What is a dilation? What is a similarity transformation?

   A dilation (or change of scale) is a transformation of the plane with the property that the ratio of the distance between the image of a point and a fixed point (called the "center" of the dilation) to the distance between the original point and the center is equal to a constant (called the "scale factor" of the dilation.) Moreover, the original point, its image, and the center of the dilation are collinear, with the point and its image on one side of the center. (For our purposes, the center cannot be between the point and its image.) The center of the dilation is fixed; the dilation does not move it.

   In other words, consider the dilation with a scale factor k, centered at P. Let A' denote the image of A under the dilation and B' the image of B. Then $\frac{A'P}{AP} = \frac{B'P}{BP} = k$, or $A'P = k(AP)$ and $B'P = k(BP)$. In the picture below, $k = 3$. That's because $A'P$ is three times as long as AP. Similarly, $B'P = 3(BP)$. Notice that because $k > 1$, the points move away from P after the dilation.

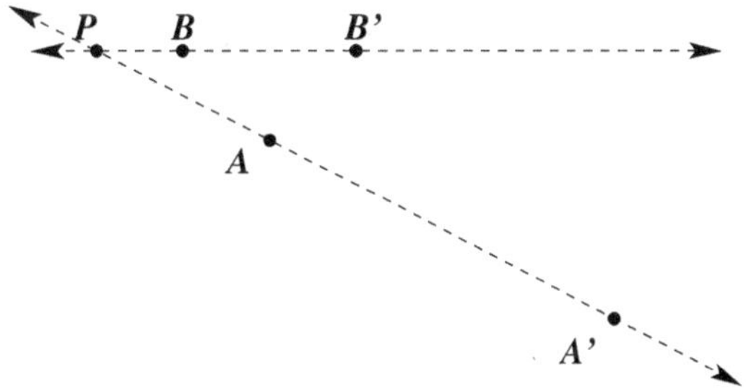

   By the way, I use the term "dilation" if $k > 1$ because then points seem to be moving away from the center. If $k < 1$, I would call it a "contraction" because points are moving toward the center, with distances getting smaller, but "dilation" has become rather common for both types, and so I will follow that usage from here on.

- What is a similarity transformation?
  A dilation is an example of a similarity transformation. More generally, a similarity transformation is a transformation that may be composed of dilations and isometries, done one after the other.
- How do you verify properties of similarity transformations experimentally?
  Probably the most precise way is to play around with geometry software. It can conduct these dilations for you and you can use the software to measure various distances and to construct ratios. However, you can also just get out pencil and paper and a ruler to experiment with dilations. We will draw pictures for each of the properties above, which provide some evidence as to why they are true.
- Basic properties of similarity transformations
  i. A dilation takes a line not passing through the center of the dilation to a parallel line, and leaves a line passing through the center unchanged.

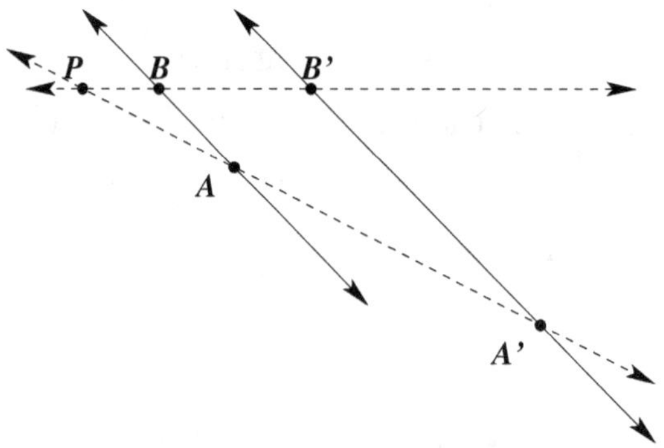

Notice that $\overleftrightarrow{A'B'}$ appears to be parallel to $\overleftrightarrow{AB}$. Moreover, lines through P are not changed: $\overleftrightarrow{A'P}$ is the same line as $\overleftrightarrow{AP}$; $\overleftrightarrow{B'P}$ is the same line as $\overleftrightarrow{BP}$.

  ii. The dilation of a line segment is longer or shorter in the ratio given by the scale factor.

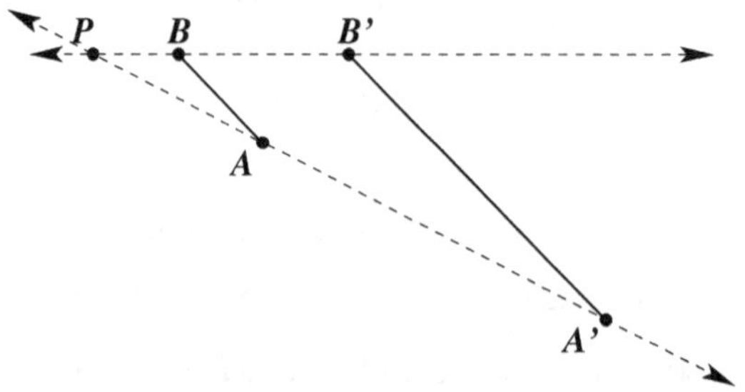

We know from how the dilation is defined that $\frac{A'P}{AP} = \frac{B'P}{BP} = k$. (In this picture, $k = 3$.) But if you draw precisely and measure closely, you will find that $\frac{A'B'}{AB}$ also equals 3. In general, any line segment $\overline{ST}$ will transform under a dilation with scale factor k to $\overline{S'T'}$ with $S'T' = k(ST)$.

iii. A dilation preserves angles.

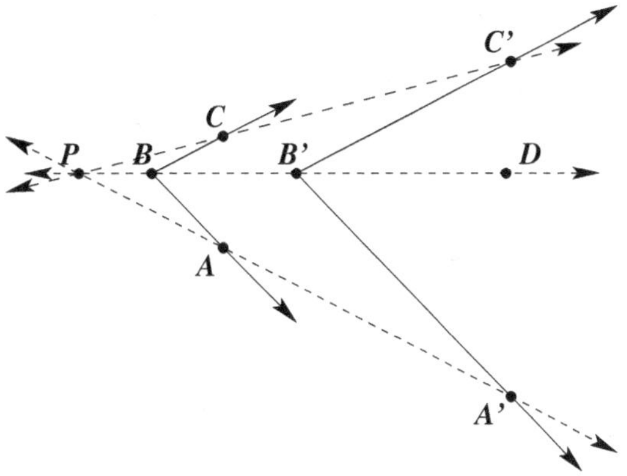

Here, $\angle ABC$ is congruent to $\angle A'B'C'$. Unlike the previous two properties, we can actually prove this one using parallel lines and transversals, and assuming that the first property above is true.

To see this, notice that $\angle ABB' \cong \angle A'B'D$ because they are corresponding angles when parallel lines $\overleftrightarrow{AB}$ and $\overleftrightarrow{A'B'}$ are cut by transversal $\overleftrightarrow{PD}$. Similarly, $\angle CBB' \cong \angle C'B'D$. So, by what is sometimes called the Angle Addition Property, we can deduce that $\angle ABC$ (made up of $\angle ABB'$ and $\angle CBB'$) must be congruent to $\angle A'B'C'$ (made up of $\angle A'B'D$ and $\angle C'B'D$, which are congruent, respectively, to $\angle ABB'$ and $\angle CBB'$).

– Why do we need to verify properties of similarity transformations experimentally? Why can't we just prove them?

Remember that there are undefined terms in geometry, like point and line. The reason for that is that you have to start somewhere without a definition so that subsequent terms can be defined in terms of those words.

A similar situation occurs in a logical system, like geometry. You can't just prove everything. A proof relies on other facts and other statements. Some of those statements must just be assumed to be true so that there is a starting point. Such statements are called "axioms" or "postulates."

Often the topic of similarity in geometry is introduced using a postulate that two triangles are similar if two pairs of their corresponding angles are congruent. All the subsequent properties of similarity, including the properties of similarity transformations that we just examined and verified experimentally, can then be proved as logical consequences of this postulate.

The Common Core State Standards explain similarity based on transformational geometry, starting with similarity transformations. Then two triangles (or other figures) will be defined (below) to be similar if one is the image of the other under a similarity transformation. This is our starting point. Then, we will deduce the other properties of similarity from this starting point. These properties that we just verified, then, are like the postulates of our system. One advantage of using transformational geometry as the basis of your logical system is that you can verify its postulates experimentally. By measuring and drawing pictures, you can convince yourself that the postulates really should be true.

– Which properties of dilations are also true of similarity transformations in general?

Recall that a similarity transformation is composed of dilations and isometries. This means that the first property is not necessarily true of a general similarity transformation. If you rotate or reflect a line, its image does not have to be parallel to the original line.

However, an isometry does not change lengths of segments. So, composing dilations and isometries still means that each line segment's length will be changed by the same scale factor, namely the product of all the scale factors of the individual dilations that make up the similarity transformation.

Finally, angles are still preserved. Since every dilation and isometry preserves angles, any combination of them must also preserve angles.

2. Given two figures, use the definition of similarity in terms of similarity transformations to decide if they are similar; explain using similarity transformations the meaning of similarity for triangles as the equality of all corresponding pairs of angles and the proportionality of all corresponding pairs of sides.

   – What is the definition of "similar"? How is it notated?

   Two geometric objects are similar if one of them is the image of the other under a similarity transformation. If $\triangle ABC$ and $\triangle DEF$ are similar, then we write $\triangle ABC \sim \triangle DEF$. As with congruence, the order of listing the points is important. In this example, the similarity transformation would map A to D, B to E

and C to F.

- How can you tell if two figures are similar?

  You have to determine if one figure is a scaled version of the other figure. That means all ratios of corresponding line segment lengths have to be equal. Also, any corresponding angles must be congruent.

- What are some consequences of two triangles being similar?

  Suppose $\triangle ABC \sim \triangle DEF$. From our properties above, that means that $\angle A \cong \angle D; \angle B \cong \angle E$; and $\angle C \cong \angle F$, because angles are preserved. Also, it follows that there is a value of $k > 0$ so that $DE = k(AB); DF = k(AC); EF = k(BC)$. So, the definition of two triangles being similar means that all corresponding pairs of angles are congruent and all corresponding pairs of sides are in the same proportion.

3. Use the properties of similarity transformations to establish the AA criterion for two triangles to be similar.

- What is the AA criterion?

  The AA criterion says that two triangles are similar if they have two pairs of corresponding angles that are congruent. Notice that this could also be called AAA, because if you know two angles of a triangle, then you can determine the third. So, if two triangles share two angles in common, then all three angles of one triangle are respectively congruent to the three angles of the other triangle.

- How do we deduce the AA criterion from the properties of similarity transformations?

  Suppose that in $\triangle ABC$ and $\triangle DEF$, we have $\angle A \cong \angle D$ and $\angle B \cong \angle E$. We must show that the triangles are similar. Consider side lengths of $\overline{AB}$ and $\overline{DE}$. Let $k = \dfrac{DE}{AB}$. Perform a dilation on $\triangle ABC$ (centered at A, perhaps, as in the picture, though the center doesn't matter) with scale factor k. Let $\triangle A'B'C'$ denote the image of $\triangle ABC$ under this dilation.

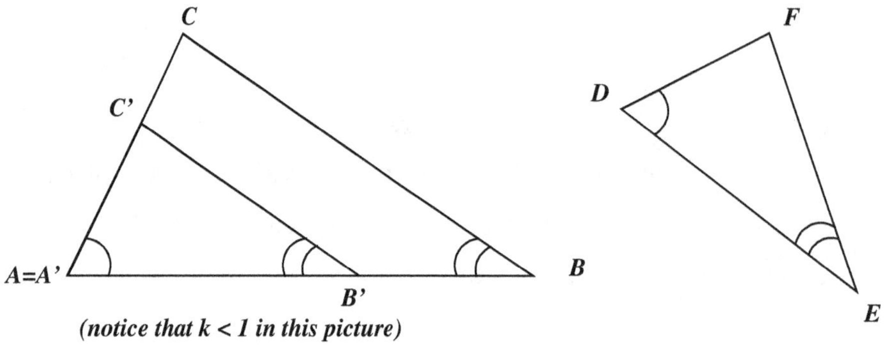

(notice that k < 1 in this picture)

Since dilations preserve angles, $\triangle A'B'C'$ has the same three angles as $\triangle ABC$, but now $A'B' = k(AB) = DE$. So $\overline{A'B'} \cong \overline{DE}$. Therefore, by ASA, $\triangle A'B'C' \cong \triangle DEF$. This means that there is a rigid motion that will take $\triangle A'B'C'$ to $\triangle DEF$. So, if we compose the above dilation with the rigid motion, we have found a similarity transformation that takes $\triangle ABC$ to $\triangle DEF$. Therefore $\triangle ABC \sim \triangle DEF$. This completes the proof.

- PROVE THEOREMS INVOLVING SIMILARITY

4. **Prove theorems about triangles.** *Theorems include: a line parallel to one side of a triangle divides the other two proportionally, and conversely; the Pythagorean Theorem proved using triangle similarity.*

   – How do you prove that a line parallel to one side of a triangle divides the other two sides proportionally?
   In the picture below, suppose that $\overline{MN}$ is parallel to $\overline{BC}$.

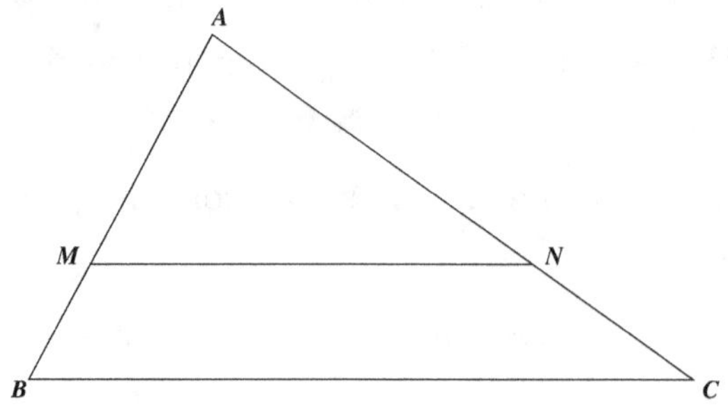

Viewing $\overline{AB}$ as a transversal, we see that $\angle AMN \cong \angle ABC$ because they are corresponding angles. Clearly $\angle A \cong \angle A$. So, by AA, $\triangle AMN \sim \triangle ABC$. This means there exists some scale factor k satisfying $k(AM) = AB$ and $k(AN) = AC$.

Notice that $AB = AM + MB$ and $AC = AN + NC$. A little algebra yields:

$$\begin{aligned} k = \frac{AB}{AM} &= \frac{AC}{AN} \\ \frac{AM + MB}{AM} &= \frac{AN + NC}{AN} \\ 1 + \frac{MB}{AM} &= 1 + \frac{NC}{AN}, \end{aligned}$$

from which it follows that $\frac{MB}{AM} = \frac{NC}{AN}$. Therefore, $\overline{MN}$ divides sides $\overline{AB}$ and $\overline{AC}$ proportionally. This completes the proof.

- How do you prove that if a line divides two sides of a triangle proportionally, then it is parallel to the third side?

  In the picture below, suppose that $\frac{MB}{AM} = \frac{NC}{AN}$. Prove that $\overline{MN}$ is parallel to $\overline{BC}$.

  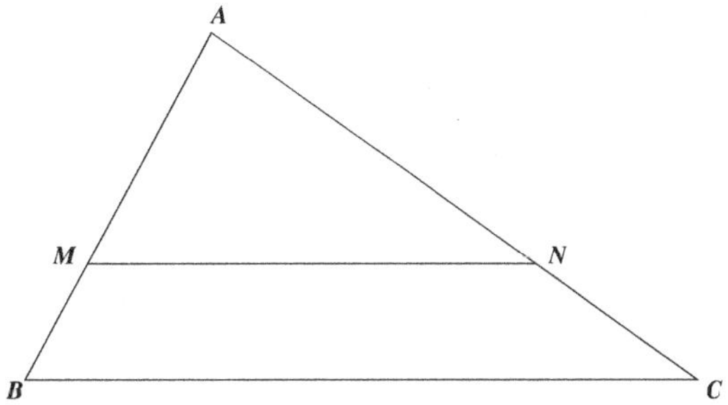

  Following similar algebra as above, in reverse, we find that there exists a scale factor k that satisfies $k(AM) = AB$ and $k(AN) = AC$. Consider what happens when $\triangle AMN$ experiences a dilation centered at A with scale factor k. Point A is fixed under this dilation. The image of M under this dilation is B because A, B, and M are collinear, and $k(AM) = AB$. Similarly, the image of N is C. Thus, the image of $\triangle AMN$ under the dilation is $\triangle ABC$. By definition of similar, $\triangle AMN \sim \triangle ABC$. Hence, $\angle AMN \cong \angle ABC$, because corresponding angles of similar triangles are congruent. But $\angle AMN$ and $\angle ABC$ are corresponding angles when $\overline{MN}$ and $\overline{BC}$ are cut by transversal $\overline{AB}$. Since these angles are congruent, $\overline{MN}$ is parallel to $\overline{BC}$. This completes the proof.

- What triangles are similar in a right triangle with the altitude to the hypotenuse drawn? Why are these useful?

  Consider the picture below, in which $\triangle ABC$ has a right angle at C, and altitude $\overline{CD}$ is drawn. By the definition of altitude, $\angle ADC$ and $\angle BDC$ are right angles. So, $\angle ADC \cong \angle ACB$. Also, $\angle A \cong \angle A$. Thus, by AA, $\triangle ACB \sim \triangle ADC$. The same line of reasoning leads to $\triangle CDB \sim \triangle ACB$. By transitivity of similarity, then, we also have $\triangle ADC \sim \triangle CDB$.

These similar triangles are useful in applications, in trigonometry, and in that they provide a way to prove the Pythagorean Theorem.

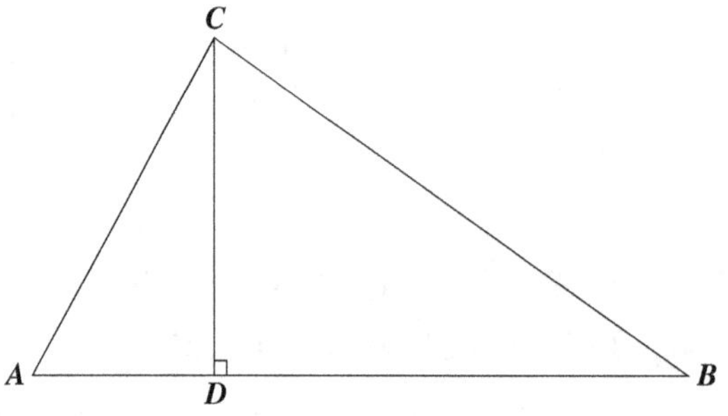

– What does the Pythagorean Theorem say?

If $\triangle ABC$ has a right angle at C, then $a^2 + b^2 = c^2$, where a refers to BC, the length of the side opposite angle A. There is a similar definition for b and c.

– How do you prove the Pythagorean Theorem using similar triangles?

Using the picture above, let $a = BC$, $b = AC$, $c = AB$, $x = AD$, and $y = BD$. [Notice that $x + y = c$.] Recall that $\triangle ACB \sim \triangle CDB \sim \triangle ADC$. So, $\frac{c}{a} = \frac{a}{y}$ and $\frac{c}{b} = \frac{b}{x}$. After cross-multiplying, we get $a^2 = cy$ and $b^2 = cx$. Thus

$$a^2 + b^2 = cy + cx = c(y + x) = c^2.$$

5. Use congruence and similarity criteria for triangles to solve problems and to prove relationships in geometric figures.

– What does "CPCTC" mean?

In my geometry class, this was our abbreviation for "corresponding parts of congruent triangles are congruent." It is most helpful in proofs, and will be occasionally used in what follows.

– Sample Problems (Answers are given below.)

(a) Complete the statement and then prove it: In parallelogram $ABCD$, $\triangle ABC \cong \triangle$__________.

(b) Suppose that X is the common midpoint of $\overline{AB}$ and $\overline{CD}$. Then $\triangle AXC \cong \triangle$__________. Prove it.

(c) The Isosceles Triangle Theorem says that if two sides of a triangle are congruent, then the angles opposite those sides are also congruent. Prove the Isosceles Triangle Theorem by drawing the angle bisector of the vertex angle.

(d) Prove that the diagonals of a rhombus bisect their vertex angles.

(e) True or false: Any two isosceles triangles must be similar. If true, prove it, and if false, give a counterexample.

(f) Draw an isosceles trapezoid and its two diagonals. The trapezoid is now divided into four triangular regions. Prove that two of these regions must be similar. Then prove that the other two regions must be congruent.

(g) Sketch square $ABCD$ and its image under a dilation of scale factor 2 ...

i. centered at the center of the square.

ii. centered at point A.

iii. centered at the midpoint of $\overline{CD}$.

(h) Consider $\triangle ABC$ with $a = 3$, $b = 4$, and $c = 5$. Draw altitude $\overline{CD}$. Find the scale factor of the dilation that maps ...

i. $\triangle ABC$ to $\triangle CBD$.

ii. $\triangle ABC$ to $\triangle ACD$.

(i) Is the composition of two similarity transformations another similarity transformation? Explain. If so, what is its scale factor?

(j) Is the composition of a similarity transformation and an isometry another similarity transformation? Explain. If so, what is its scale factor?

(k) Join the midpoints of the sides of a square to obtain a smaller square. What is the exact similarity transformation that maps the first square to the second square? Is there more than one right answer?

(l) What scale factor is needed for a similarity transformation to transform a square into a square with twice the area?

– Answers to Sample Problems

(a) Complete the statement and then prove it: In parallelogram $ABCD$, $\triangle ABC \cong \underline{\triangle CDA}$.

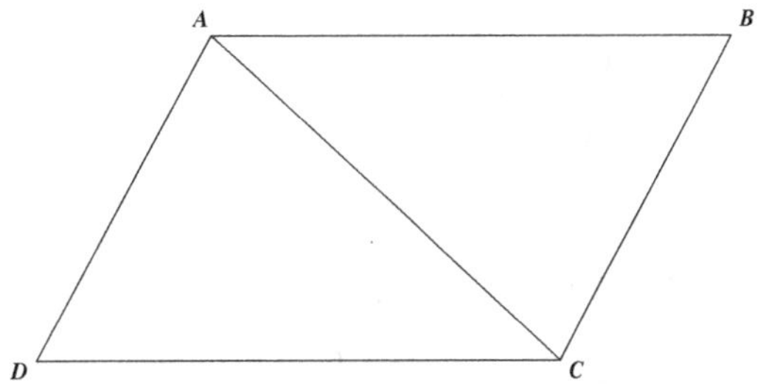

Proof: Draw diagonal $\overline{AC}$. Since $\overline{AB} \parallel \overline{CD}$, $\angle BAC \cong \angle DCA$. Also, since $\overline{AD} \parallel \overline{BC}$, $\angle BCA \cong \angle DAC$. And of course $\overline{AC} \cong \overline{AC}$. Therefore, by ASA, $\triangle ABC \cong \triangle CDA$. □

(b) Suppose that X is the common midpoint of $\overline{AB}$ and $\overline{CD}$. Then $\triangle AXC \cong \triangle BXD$. Prove it.

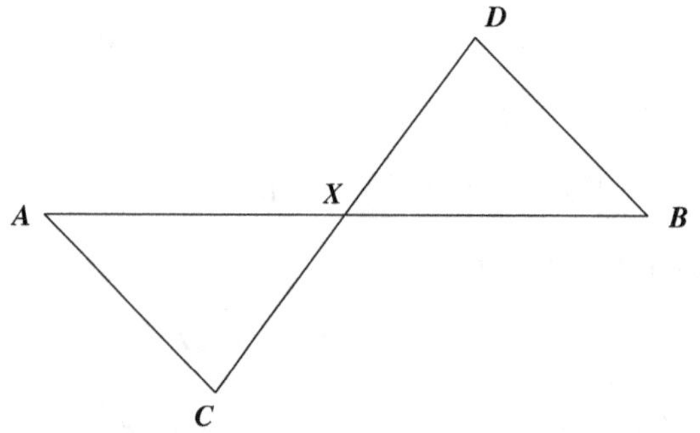

Proof: Draw $\overline{AC}$ and $\overline{BD}$. Since X is the midpoint of $\overline{AB}$, $\overline{AX} \cong \overline{XB}$. Similarly, $\overline{CX} \cong \overline{XD}$. Also, $\angle AXC \cong \angle BXD$ because they are vertical. Therefore, by SAS, $\triangle AXC \cong \triangle BXD$. □

(c) The Isosceles Triangle Theorem says that if two sides of a triangle are congruent, then the angles opposite those sides are also congruent. Prove the Isosceles Triangle Theorem by drawing the angle bisector of the vertex angle. Given: In $\triangle ABC$, $\overline{AB} \cong \overline{AC}$. Prove that $\angle B \cong \angle C$.

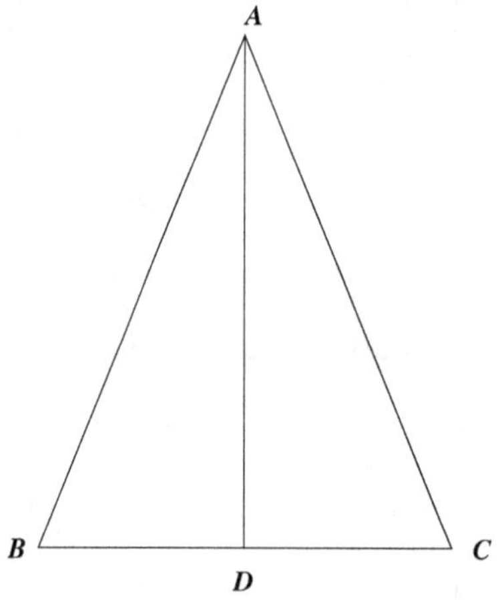

Proof: Draw $\overline{AD}$, the angle bisector of $\angle A$. This means that $\angle DAB \cong \angle DAC$. We are told that $\overline{AB} \cong \overline{AC}$, and clearly $\overline{AD} \cong \overline{AD}$. So, by SAS, $\triangle DAB \cong \triangle DAC$. Therefore, $\angle B \cong \angle C$ because CPCTC (corresponding parts of congruent triangles are congruent).

(d) Prove that the diagonals of a rhombus bisect their vertex angles.

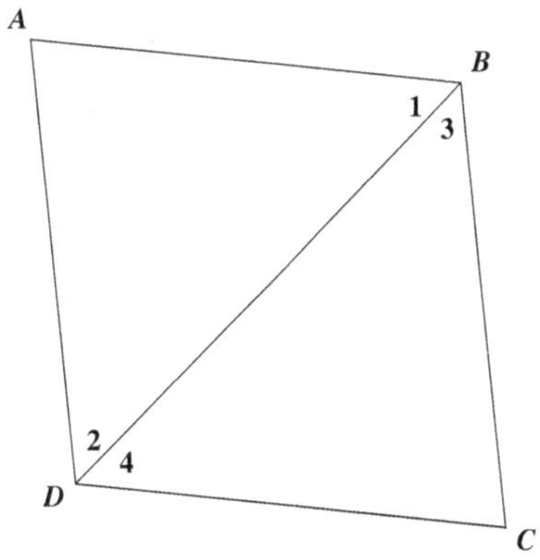

Proof: Since $ABCD$ is a rhombus, we know $\overline{AB} \cong \overline{BC} \cong \overline{CD} \cong \overline{DA}$ and clearly, $\overline{BD} \cong \overline{BD}$. Thus, by SSS, $\triangle ABD \cong \triangle CBD$. So, $\angle 1 \cong \angle 3$ and $\angle 2 \cong \angle 4$ by CPCTC. Therefore, $\overline{BD}$ bisects $\angle ABC$ and $\angle CDA$. Because we used arbitrary vertices in an arbitrary rhombus, our proof also shows that the other diagonal bisects its vertex angles as well. □

(e) True or false: Any two isosceles triangles must be similar. If true, prove it, and if false, give a counterexample. FALSE

Consider an isosceles triangle with a third side shorter than the other two. (See the picture used in the Isosceles Triangle proof, above.) This happens when the vertex angle is less than 60 degrees, because a vertex angle of 60 degrees would make our isosceles triangle into an equilateral triangle. Now consider the isosceles triangle with side lengths 1, 1, and $\sqrt{2}$. Then the third side is longer than the other two. (This happens to be a right triangle; its vertex angle is 90 degrees.) Since similar triangles must have congruent angles, there is no way that these two isosceles triangles can be similar.

(f) Draw an isosceles trapezoid and its two diagonals. The trapezoid is now divided into four triangular regions. Prove that two of these regions must be similar. Then prove that the other two regions must be congruent.

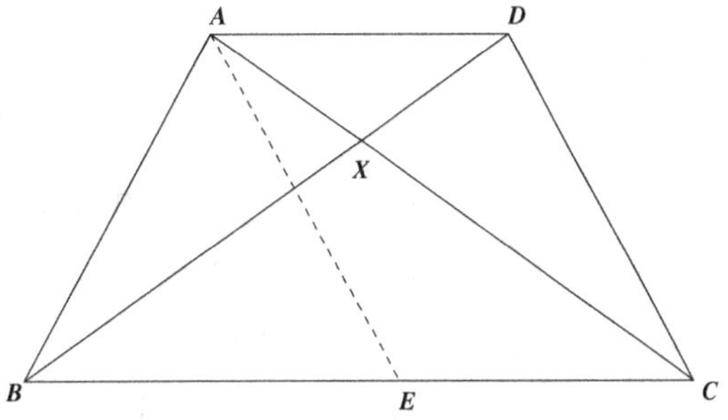

Proof: We are given isosceles trapezoid $ABCD$, in which $\overline{AD} \parallel \overline{BC}$ and

$\overline{AB} \cong \overline{CD}$. Since alternate interior angles are congruent whenever parallel lines are cut by a transversal, $\angle DAC \cong \angle BCA$. Also, $\angle AXD \cong \angle CXB$ because they are vertical angles. So, by AA, $\triangle AXD \sim \triangle CXB$. □

Now, draw $\overline{AE}$ parallel to $\overline{CD}$. Thus, $AECD$ is a parallelogram, and hence, $\overline{AE} \cong \overline{CD}$. By the transitive property of congruence, $\overline{AE} \cong \overline{AB}$. This makes $\triangle AEB$ isosceles, and so by the Isosceles Triangle Theorem, $\angle ABE \cong \angle AEB$. Because they are corresponding angles, $\angle AEB \cong \angle DCB$. By transitivity, then, $\angle ABC \cong \angle DCB$. Since $\overline{BC} \cong \overline{BC}$, SAS implies that $\triangle ABC \cong \triangle DCB$. Thus, $\angle BAC \cong \angle CDB$ by CPCTC. Moreover, $\angle AXB \cong \angle DXC$ because they are vertical. Therefore, by AAS, $\triangle AXB \cong \triangle DXC$. □

(g) Sketch square $ABCD$ and its image under a dilation of scale factor 2 ...

i. centered at the center of the square.

ii. centered at point A.

iii. centered at the midpoint of $\overline{CD}$.

In each picture, $A'B'C'D'$ is the image of $ABCD$ after the transformation.

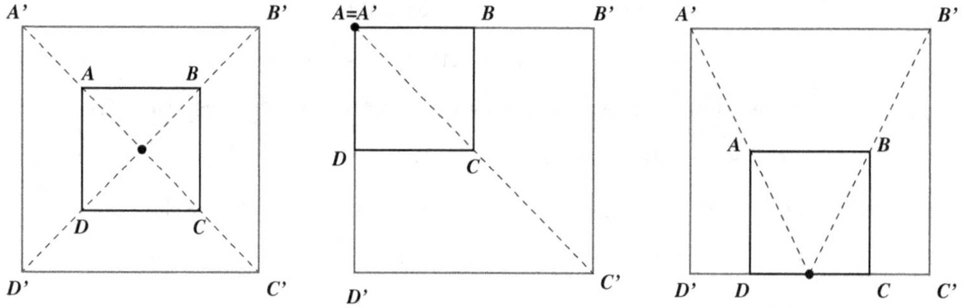

(h) Consider $\triangle ABC$ with $a = 3$, $b = 4$, and $c = 5$. Draw altitude $\overline{CD}$. Find the scale factor of the dilation that maps ...

i. $\triangle ABC$ to $\triangle CBD$. $\frac{3}{5}$

ii. $\triangle ABC$ to $\triangle ACD$. $\frac{4}{5}$

(i) Is the composition of two similarity transformations another similarity transformation? Explain. If so, what is its scale factor?

The composition of two similarity transformations is a similarity transformation and the scale factor of the composition is the product of the scale factors of the individual similarity transformations. If S_1 is a similarity transformation with scale factor k_1 and S_2 is a similarity transformation with scale factor k_2, then the composition of S_1 and S_2 is a similarity transformation with scale factor $k_1 k_2$.

(j) Is the composition of a similarity transformation and an isometry another similarity transformation? Explain. If so, what is its scale factor?
The composition of a similarity transformation and an isometry is another similarity transformation. Moreover, the composition has a scale factor equal to the scale factor of the first similarity transformation. If S is a similarity transformation with scale factor k and T is an isometry, then the composition of S and T is a similarity transformation with scale factor k.

(k) Join the midpoints of the sides of a square to obtain a smaller square. What is the exact similarity transformation that maps the first square to the second square? Is there more than one right answer?
There are several right answers. This transformation is composed of a dilation and a rotation. The dilation is centered at the center of the square and has a scale factor of $\frac{1}{\sqrt{2}} = \frac{\sqrt{2}}{2}$. This is followed by a rotation of 45 degrees around the center of the square.

(l) What scale factor is needed for a similarity transformation to transform a square into a square with twice the area? $\sqrt{2}$

- Define trigonometric ratios and solve problems involving right triangles

6. **Understand that by similarity, side ratios in right triangles are properties of the angles in the triangle, leading to definitions of trigonometric ratios for acute angles.**

   – Why do ratios of side lengths in right triangles depend only on the angles of the triangle?
   The answer has to do with triangle similarity. We will show two facts: 1) if two right triangles have the same angles, then their side ratios are the same; and 2) if two right triangles have the same side ratios, then their angles are the same. To be more explicit, let $\triangle ABC$ and $\triangle DEF$ be right triangles, with right angles at C and F, respectively. So, $\angle C \cong \angle F$.

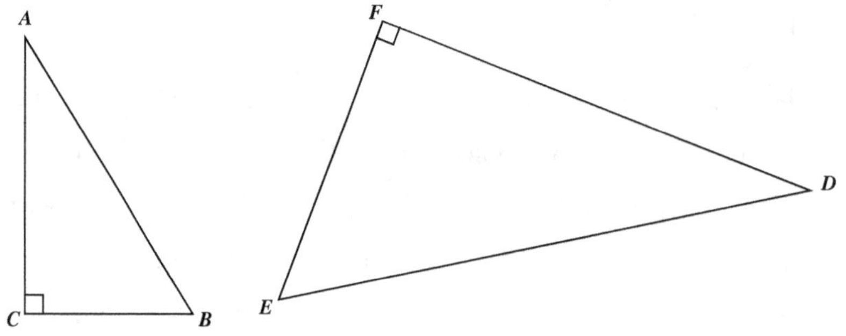

To begin, suppose that $\angle A \cong \angle D$. Then, by AA, we know that $\triangle ABC \sim \triangle DEF$. Then there exists a scale factor k satisfying: $DE = k(AB)$; $EF = k(BC)$; and $DF = k(AC)$. Solving for k in each equation gives:

$$k = \frac{DE}{AB} = \frac{EF}{BC} = \frac{DF}{AC}.$$

Algebraic manipulation of $\frac{DE}{AB} = \frac{EF}{BC}$ allows us to deduce $\frac{BC}{AB} = \frac{EF}{DE}$. So, the ratio of one leg to the hypotenuse is the same in each triangle. The other side ratios can be deduced in the same way. Therefore, if two right triangles have all the same angles, then their side ratios are the same.

Conversely, suppose that $\triangle ABC$ and $\triangle DEF$ have the same side ratios. We will show that their angles must be the same. Because the side ratios are the same, we can say that $\frac{DF}{EF} = \frac{AC}{BC}$, or, after some algebra, $\frac{DF}{AC} = \frac{EF}{BC}$. Let $k = \frac{DF}{AC} = \frac{EF}{BC}$. Then $DF = k(AC)$ and $EF = k(BC)$. Now consider what happens when $\triangle ABC$ experiences a dilation centered at C with scale factor k. Say $\triangle A'B'C'$ is the image of $\triangle ABC$ under this dilation.

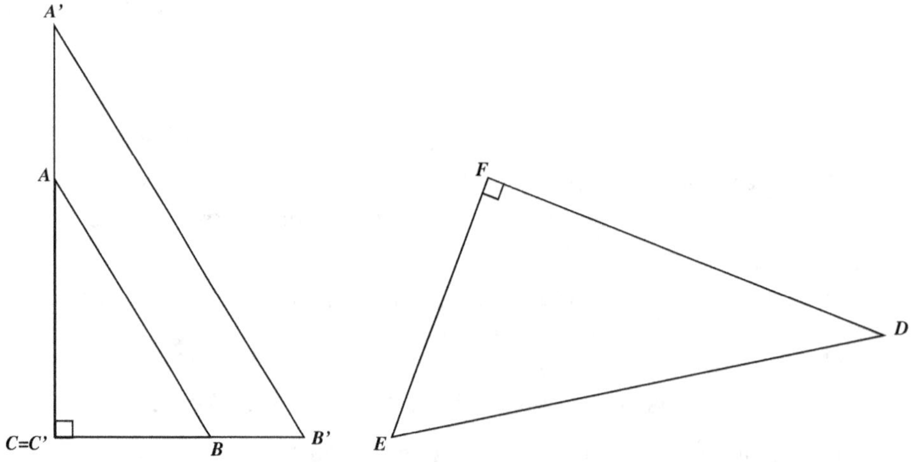

Then $A'C' = k(AC) = DF$ and $B'C' = k(BC) = EF$. So, $\overline{A'C'} \cong \overline{DF}$ and $\overline{B'C'} \cong \overline{EF}$. Moreover, since angles are not changed under a dilation, $\angle C' \cong \angle C \cong \angle F$. So, by SAS, $\triangle A'B'C' \cong \triangle DEF$. We can therefore deduce that

$\angle A \cong \angle A' \cong \angle D$ and $\angle B \cong \angle B' \cong \angle E$. Thus all three angles of $\triangle ABC$ match up with all three angles of $\triangle DEF$.

The proof is complete. In fact, these statements are really true of any two triangles, not just right triangles.

– If these two statements are true of any two triangles, then why do we need to use right triangles?

The main reason we use right triangles is because of the Pythagorean Theorem. It will play a large role in the trigonometry that follows.

Another reason we use right triangles is for standardization. If you know one acute angle in a right triangle, then you have determined all the angles of the triangle. Up to a scale factor, there is only one right triangle with an acute angle of 30 degrees, say. So, all the side ratios can be determined once you know that one of the acute angles is 30 degrees. If we were to just use any triangle with a 30 degree angle, then we would not have enough information to determine what the side ratios of that triangle would be.

– What are the standard trigonometric ratios? Which side length ratios are they?

There are six ways to form the ratio of two distinct sides of a triangle. Thus there are six distinct trigonometric ratios. The six trigonometric ratios are sine, cosine, tangent, cotangent, secant, and cosecant (csc). Usually in Geometry, only sine, cosine, and tangent are used, while the others are written in terms of these three. In what follows, the legs are labeled in reference to angle A: the opposite leg (opp) and the adjacent leg (adj). The hypotenuse is abbreviated "hyp." For simplicity, we also write $\sin A$ rather than $\sin(\angle A)$, etc.

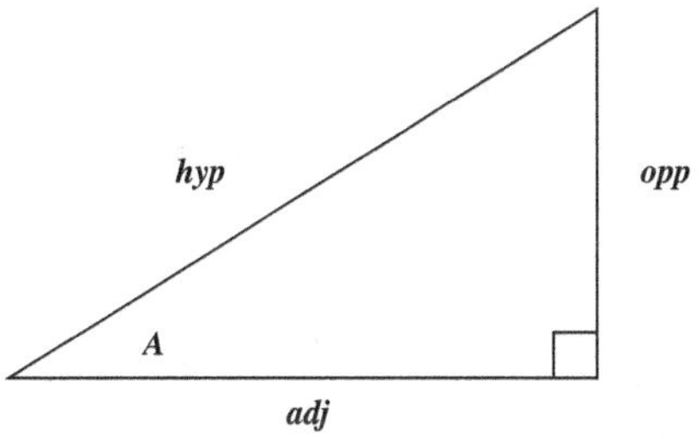

$$\sin A = \frac{opp}{hyp}; \ \cos A = \frac{adj}{hyp}; \ \tan A = \frac{opp}{adj} = \frac{\sin A}{\cos A};$$

$$\csc A = \frac{hyp}{opp} = \frac{1}{\sin A}; \ \sec A = \frac{hyp}{adj} = \frac{1}{\cos A}; \ \cot A = \frac{adj}{opp} = \frac{1}{\tan A} = \frac{\cos A}{\sin A}$$

7. Explain and use the relationship between the sine and cosine of complementary angles.

– What is the relationship between the sine and cosine of complementary angles?

Recall that complementary angles are angles whose sum is a right angle. So, in a right triangle, the two acute angles are complementary to each other. (If the sum of all three angles is 180, and one angle is 90, then the sum of the other two must also be 90.)

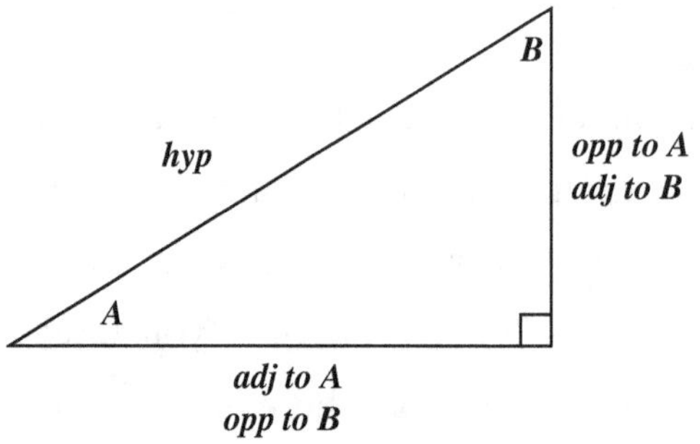

This relationship hinges on one's point of view. The leg that is opposite angle A is the leg that is adjacent to angle B. So

$$\sin A = \frac{opp(A)}{hyp} = \frac{adj(B)}{hyp} = \cos B.$$

Similarly, $\cos A = \sin B$. In general, $\cos(x) = \sin(90 - x)$.

As a way to remember this relationship, the word "cosine" itself comes from a contraction of "complement's sine." The other "co-" functions have a similar meaning.

– How do you use the relationship between the sine and cosine of complementary angles?

Using this relationship depends largely on the situation. For some examples of how it might be useful in problem solving, see the problems in the next section.

8. Use trigonometric ratios and the Pythagorean Theorem to solve right triangles in applied problems.**

– Is there a connection between the Pythagorean Theorem and trigonometric ratios? If so, what is it?

Yes, there is a connection. I like to call this connection the most important trigonometric identity there is. We will deduce the connection based on $\triangle ABC$, with a right angle at C and where a is the side opposite $\angle A$, etc.

$$a^2 + b^2 = c^2 \Leftrightarrow \frac{a^2}{c^2} + \frac{b^2}{c^2} = \frac{c^2}{c^2} \Leftrightarrow \left(\frac{a}{c}\right)^2 + \left(\frac{b}{c}\right)^2 = 1 \Leftrightarrow \sin^2 A + \cos^2 A = 1$$

So, for any angle x, $\sin^2 x + \cos^2 x = 1$. This is the Pythagorean theorem in trigonometry form.

– How do you use trigonometric ratios and the Pythagorean Theorem to solve problems?

  We'll do one example below, and then list several sample problems.

  Suppose you want to determine the height of a tall tree. You stand 50 feet away from the tree and measure the angle of elevation from your eye (which is 5 feet off the ground) to the top of the tree. The angle of elevation you get is 39 degrees. How tall is the tree?

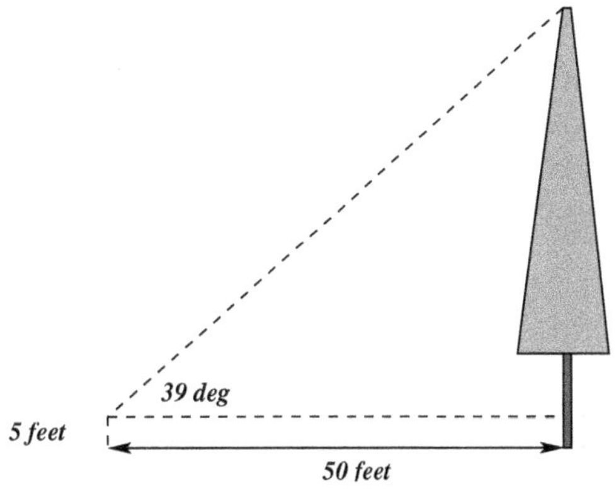

  First, we will find the side of the triangle that contains the tree. Notice that the tangent of 39 degrees is this height divided by 50. So, this height is $50 \tan 39 \approx 40.49$ feet. But this is just the height of the tree from our eye level upward. So we need to add 5 feet in order to obtain the total height of the tree. The tree is thus 45.49 feet tall, or about 45 feet and 6 inches.

– Sample Problems (Answers given below.)

  (a) Write down all trigonometric ratios for a 30-60-90 triangle and a 45-45-90 triangle.

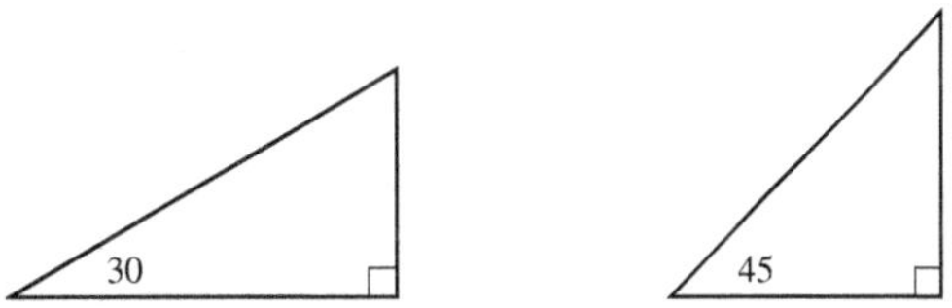

  (b) Write down all the trigonometric ratios for a right triangle with legs 8 and 15.

  (c) Complete the table without using a calculator.

θ	$\sin\theta$	$\cos\theta$
$75°$		
$15°$		$\sqrt{\frac{2+\sqrt{3}}{4}}$

(d) Can you define all the trigonometric ratios for 0 and 90 degree angles? If so, do it, and if not, why not?

(e) Explain how the Triangle Inequality follows from the saying "the shortest distance between two points is a straight line."

– Answers to Sample Problems

(a) Write down all trigonometric ratios for a 30-60-90 triangle and a 45-45-90 triangle.

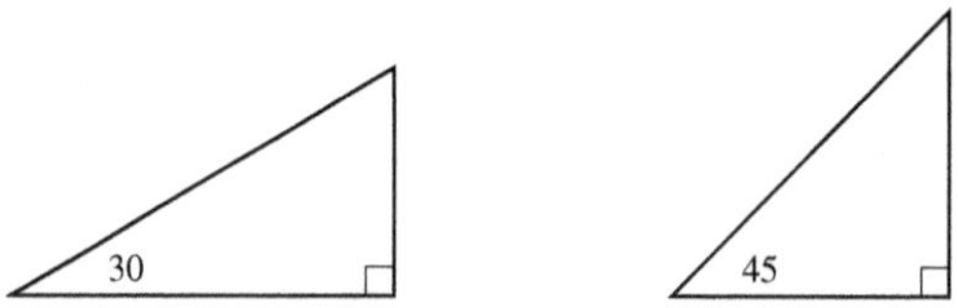

Angle θ	$\sin\theta$	$\cos\theta$	$\tan\theta$	$\cot\theta$	$\sec\theta$	$\csc\theta$
30	$\frac{1}{2}$	$\frac{\sqrt{3}}{2}$	$\frac{1}{\sqrt{3}} = \frac{\sqrt{3}}{3}$	$\sqrt{3}$	$\frac{2}{\sqrt{3}} = \frac{2\sqrt{3}}{3}$	2
60	$\frac{\sqrt{3}}{2}$	$\frac{1}{2}$	$\sqrt{3}$	$\frac{1}{\sqrt{3}} = \frac{\sqrt{3}}{3}$	2	$\frac{2}{\sqrt{3}} = \frac{2\sqrt{3}}{3}$
45	$\frac{\sqrt{2}}{2}$	$\frac{\sqrt{2}}{2}$	1	1	$\sqrt{2}$	$\sqrt{2}$

(b) Write down all the trigonometric ratios for a right triangle with legs 8 and 15. First, notice that the hypotenuse is $\sqrt{8^2 + 15^2} = \sqrt{289} = 17$.

Angle	$\sin\theta$	$\cos\theta$	$\tan\theta$	$\cot\theta$	$\sec\theta$	$\csc\theta$
opposite 8	$\frac{8}{17}$	$\frac{15}{17}$	$\frac{8}{15}$	$\frac{15}{8}$	$\frac{17}{15}$	$\frac{17}{8}$
opposite 15	$\frac{15}{17}$	$\frac{8}{17}$	$\frac{15}{8}$	$\frac{8}{15}$	$\frac{17}{8}$	$\frac{17}{15}$

(c) Complete the table without using a calculator.

θ	$\sin\theta$	$\cos\theta$
$75°$	$\sqrt{\frac{2+\sqrt{3}}{4}}$	$\sqrt{\frac{2-\sqrt{3}}{4}}$
$15°$	$\sqrt{\frac{2-\sqrt{3}}{4}}$	$\sqrt{\frac{2+\sqrt{3}}{4}}$

Okay, this one is a little tricky algebraically. How did we get this? First, notice that the angles are complementary. So the sine of one is the cosine of the other, and vice versa. So, given that $\cos(15°) = \sqrt{\frac{2+\sqrt{3}}{4}}$, we can find

$\sin(15°)$ by using the identity: $\sin^2 x + \cos^2 x = 1$. So

$$\begin{aligned}\sin^2(15°) + \cos^2(15°) &= 1 \\ \sin^2(15°) + \frac{2+\sqrt{3}}{4} &= 1 \\ \sin^2(15°) &= 1 - \frac{2+\sqrt{3}}{4} = \frac{2-\sqrt{3}}{4}.\end{aligned}$$

(d) Can you define all the trigonometric ratios for 0 and 90 degree angles? If so, do it, and if not, why not?

Think of what happens as the angle A approaches 0 degrees. Then the opposite side shrinks to zero, and the adjacent side length approaches the length of the hypotenuse. So, $\sin(0) = 0$ and $\cos(0) = 1$. Thus, $\tan(0) = 0$ and $\sec(0) = 1$, but $\cot(0)$ and $\csc(0)$ are not defined because they have denominators equal to zero. Similarly, $\sin(90) = 1$ and $\cos(90) = 0$. So $\cot(90) = 0$ and $\csc(90) = 1$, while $\tan(90)$ and $\sec(90)$ are not defined.

(e) Explain how the Triangle Inequality follows from the saying "the shortest distance between two points is a straight line."

Consider two vertices of a triangle. There are two paths along the triangle leading from one vertex to the other. The first path is along the side joining the two vertices. The other path is along the other two sides, passing through the third vertex. According to the saying, the shortest path is the first one, the straight line joining the two vertices. Therefore, the sum of the other two side lengths must be greater than the length of the third side. This is exactly what the Triangle Inequality says.

- APPLY TRIGONOMETRY TO GENERAL TRIANGLES

9. (+) Derive the formula $A = 1/2\, ab\sin(C)$ for the area of a triangle by drawing an auxiliary line from a vertex perpendicular to the opposite side.

   – How do you derive this formula for the area of a triangle?

   To derive this formula, we begin by drawing an altitude of an arbitrary triangle, say $\triangle ABC$. That is, we draw an auxiliary line from one vertex (e.g., A) that is perpendicular to that vertex's opposite side ($\overline{BC}$).

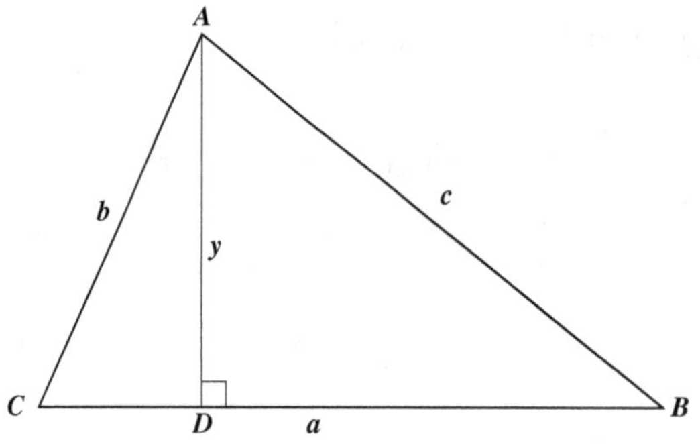

Notice that this creates right triangles, like $\triangle CDA$. By definition, $\sin C = \frac{y}{b}$, which means $y = b\sin C$. Using the formula for the area of a triangle as half the product of the base and the height, we have that the area equals $\frac{1}{2}ay = \frac{1}{2}ab\sin C$.

– Does it matter which altitude is drawn?

No, the area can be calculated using any of the three altitudes. In that case, the area is also equal to $\frac{1}{2}bc\sin A$ and $\frac{1}{2}ac\sin B$.

10. (+) Prove the Laws of Sines and Cosines and use them to solve problems.

– What does the Law of Sines say? How do you justify it?

The Law of Sines says that in $\triangle ABC$, $\frac{\sin A}{a} = \frac{\sin B}{b} = \frac{\sin C}{c}$. To justify the Law of Sines, we calculate the area of the triangle three different ways (using its three different altitudes, explained above) and then set these areas equal to each other. Thus, $\frac{1}{2}ab\sin C = \frac{1}{2}bc\sin A$. Canceling the $\frac{1}{2}b$, we get $a\sin C = c\sin A$, from which we get $\frac{\sin C}{c} = \frac{\sin A}{a}$. Using the third area formula allows you to deduce that $\frac{\sin B}{b} = \frac{\sin A}{a}$ as well.

– What does the Law of Cosines say? How do you justify it?

The Law of Cosines says that in $\triangle ABC$, $c^2 = a^2 + b^2 - 2ab(\cos C)$. To justify the Law of Cosines, we will use the following picture and a little algebra. In the picture, $a = x + y$.

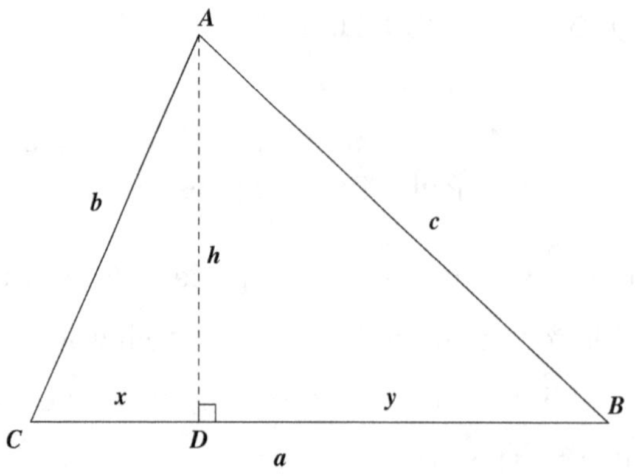

Notice that in right triangle ADC, $\cos C = \frac{x}{b}$. Thus $x = b\cos C$. Using the Pythagorean Theorem, we get $c^2 = y^2 + h^2$. So,

$$\begin{aligned} c^2 &= y^2 + h^2 = (a-x)^2 + h^2 = a^2 - 2ax + x^2 + h^2 \\ &= a^2 + (x^2 + h^2) - 2ax = a^2 + b^2 - 2ax, \end{aligned}$$

where we have applied the Pythagorean Theorem to $\triangle ADC$ to get $x^2 + h^2 = b^2$. Therefore, by substitution, $c^2 = a^2 + b^2 - 2ab\cos C$.

– Does the Law of Cosines work with other angles and sides of the triangle?
Yes, provided that the relative positions of the sides are the same. So, other forms of the Law of Cosines are:

$$a^2 = b^2 + c^2 - 2bc\cos A \text{ and } b^2 = a^2 + c^2 - 2ac\cos B.$$

– How do you use the Law of Sines and the Law of Cosines to solve problems?
These two "Laws" are most helpful when you know partial information in a triangle (not necessarily a right triangle) and you are trying to determine the rest of the information. For specific problems, see the next section.

11. (+) Understand and apply the Law of Sines and the Law of Cosines to find unknown measurements in right and non-right triangles (e.g., surveying problems, resultant forces).

– Sample Problems (Answers given below.)

(a) What happens if you apply the Law of Cosines to a triangle that has a right angle at C?

(b) Find the remaining sides of $\triangle DEF$ if $d = 4$, $m\angle E = 40$, and $m\angle F = 60$. (This is called *solving* the triangle.)

(c) Suppose that you wish to hang a welcome sign in your doorway. Gravity is pulling downward on the sign with a force of 2.4 pounds. You want to attach two strings in each upper corner of the doorway so that the strings make an angle of 20 degrees with vertical. How much force does each string need to support?

(d) Johnny wants to know how far it is to cross a river. He climbs a 10 foot high hill near the river and measures the angle of declination to each shore of the river. The angle of declination to the closer bank is 38 degrees, while the angle of declination to the far bank is 9 degrees. How far is it from where

Johnny is standing to the closer shore? How far is it from where Johnny is standing to the farther shore? How far is it across the river?

(e) Find the remaining sides of $\triangle DEF$ if $d = 13$, $m\angle E = 106$, and $m\angle F = 35$.

– Answers to Sample Problems

(a) What happens if you apply the Law of Cosines to a triangle that has a right angle at C?
The Law of Cosines says that in $\triangle ABC$, $c^2 = a^2 + b^2 - 2ab(\cos C)$. So, if $\angle C$ is a right angle, then $\cos C = 0$. Therefore, $c^2 = a^2 + b^2$, which is really the Pythagorean Theorem. Therefore, the Law of Cosines is a generalization of the Pythagorean Theorem to any triangle.

(b) Find the remaining sides of $\triangle DEF$ if $d = 4$, $m\angle E = 40$, and $m\angle F = 60$. (This is called *solving* the triangle.)
Since the angle sum of a triangle is 180, $m\angle D = 80$. Using the Law of Sines, we have

$$\frac{\sin 60}{f} = \frac{\sin 40}{e} = \frac{\sin 80}{4} \approx 0.2462.$$

Therefore, $e \approx 2.61$ and $f \approx 3.52$.

(c) Suppose that you wish to hang a welcome sign in your doorway. Gravity is pulling downward on the sign with a force of 2.4 pounds. You want to attach two strings in each upper corner of the doorway so that the strings make an angle of 20 degrees with vertical. How much force does each string need to support?
Since the sign is presumably going to hang in the doorway without moving, we need the net forces on the sign to be zero. (If the sign had a net force on it, it would be moving.) That means that the vector forces of the two strings and gravity can be thought of as three sides of a triangle, with the magnitude of each force equal to the length of its corresponding side of the triangle. So we obtain a picture like the one below.

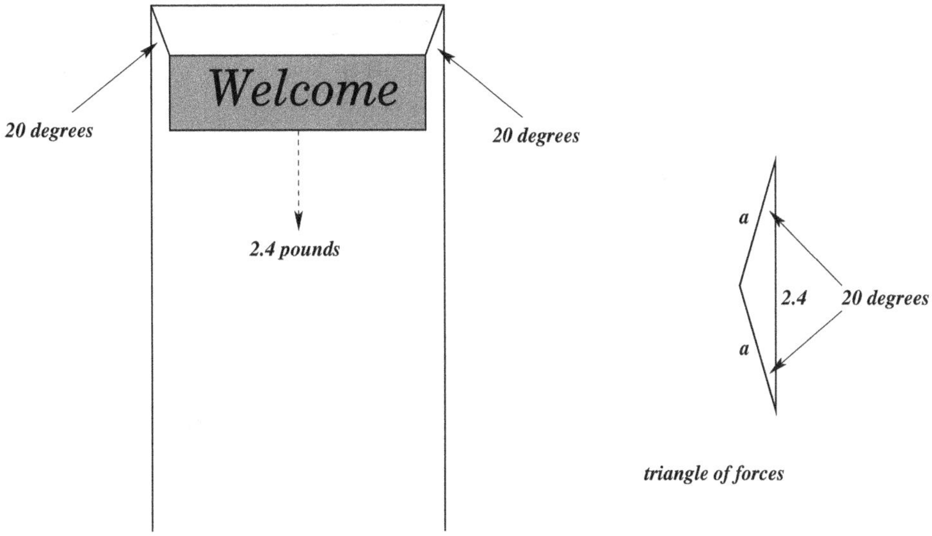

Since we know the two acute angles are 20 degrees each, then we know that the obtuse angle is 140 degrees and the vertical side's length is 2.4. Using the Law of Sines, we have $\frac{\sin 20}{a} = \frac{\sin 140}{2.4}$, or $a \approx 1.28$. So each string will need to exert about 1.28 pounds of force on the sign to hold it up.

(d) Johnny wants to know how far it is to cross a river. He climbs a hill near the river so that his eye level is 10 feet above the river, and measures the angle of declination to each shore of the river. The angle of declination to the closer bank is 38 degrees, while the angle of declination to the far bank is 9 degrees. How far is it from where Johnny is standing to the closer shore? How far is it from where Johnny is standing to the farther shore? How far is it across the river?

There are many valid solution methods to this problem. We will solve this problem in stages because each stage will help us figure out how far it is across the river. First, we sketch a picture of the situation.

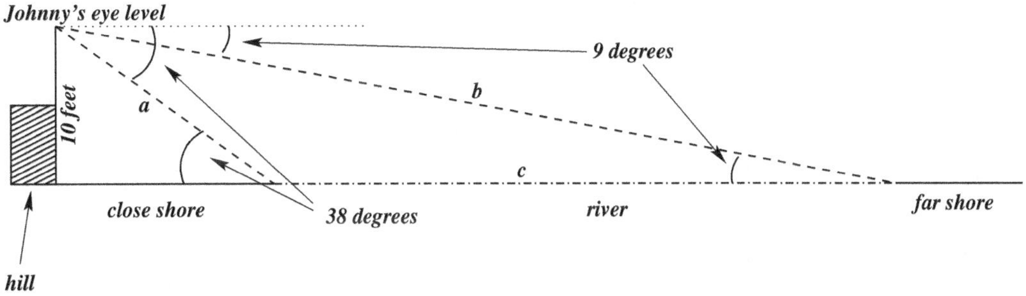

From the picture, we see that the distance from Johnny to the nearer shore, a, satisfies a trigonometric equation involving the sine of 38 degrees: $\sin 38 = \frac{10}{a}$. So, $a = \frac{10}{\sin 38} \approx 16.24$ feet. Similarly, Johnny's distance to the farther shore, b, satisfies $\sin 9 = \frac{10}{b}$. So, $b = \frac{10}{\sin 9} \approx 63.92$ feet. The distance across the river, c, can now be found using the Law of Cosines, because we know a, b, and the measure of angle C between them.

$$\begin{aligned} c^2 &= a^2 + b^2 - 2ab(\cos C) \\ &\approx (16.24)^2 + (63.92)^2 - 2(16.24)(63.92)(\cos(38-9)) \\ &\approx 263.74 + 4085.77 - 2076.12\cos(29) \\ &= 2533.69, \end{aligned}$$

from which we get $c \approx 50.33$ feet.

(e) Find the remaining sides of $\triangle DEF$ if $d = 13$, $m\angle E = 106$, and $m\angle F = 35$. We start by calculating $m\angle D = 180 - 106 - 35 = 39$ degrees. Now, using the Law of Sines, we get:

$$\frac{\sin 39}{13} = \frac{\sin 106}{e} = \frac{\sin 35}{f}.$$

Thus $e = 13\left(\frac{\sin 106}{\sin 39}\right) \approx 19.86$ and $f = 13\left(\frac{\sin 35}{\sin 39}\right) \approx 11.85$.

CIRCLES

- UNDERSTAND AND APPLY THEOREMS ABOUT CIRCLES

1. Prove that all circles are similar.

   – What does it mean to say that all circles are similar?

   By definition, a circle is the set of points that are given distance (called the "radius") from a given point (called the "center"). Recall that two objects are similar if one is taken to the other by a similarity transformation. So to say that all circles are similar means that any circle can be transformed to any other circle via a similarity transformation.

   By the way, the word "radius" also refers to any line segment from the center to the circle itself. It's a little confusing that a radius can be both a line segment and the length of that line segment, but usually the context makes things clear.

   – How do you prove that all circles are similar?

   To show that any two circles are similar, we need to start with two arbitrary circles, and show that there exists a similarity transformation taking one circle to the other. So, suppose we have two circles: circle A with radius AB, and circle C with radius CD. Set $k = \frac{CD}{AB}$. Then $CD = k(AB)$. Consider what happens to circle A under a dilation centered at A with scale factor k. Suppose A' and B' are the images of A and B under this dilation. Note that $A' = A$. Also, every radius of circle A would become a segment of length $k(AB)$. So the image of circle A with radius AB would be circle A' with radius $A'B' = k(AB)$.

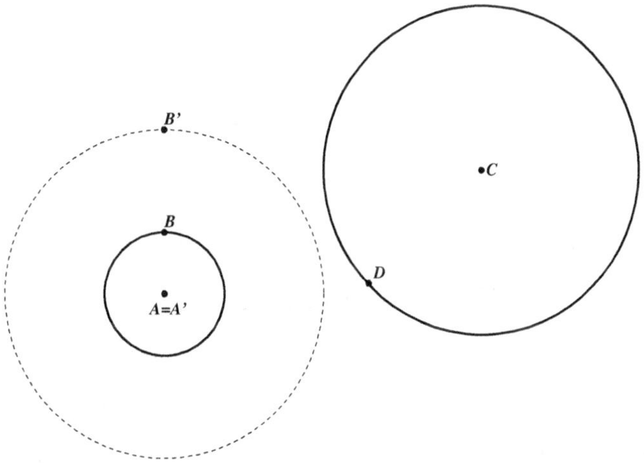

So now, let us translate circle A' so that A' moves to C. Then circle A' coincides with circle C because $A'B' = k(AB) = CD$. Composing the dilation with the translation gives us a similarity transformation that maps circle A exactly onto circle C. Therefore circle A is similar to circle C. Since these circles were completely arbitrary, any two circles are similar.

2. Identify and describe relationships among inscribed angles, radii, and chords. *Include the relationship between central, inscribed, and circumscribed angles; inscribed angles on a diameter are right angles; the radius of a circle is perpendicular to the tangent where the radius intersects the circle.*

   – What is a chord?

   A chord of a circle is a line segment whose endpoints lie on the circle. A chord is called a diameter if it passes through the center of the circle.

   – What is an inscribed angle? ... a central angle? ... a circumscribed angle?

   An angle is inscribed in a circle if its vertex lies on the circle and its two sides contain chords of the circle.

   An angle in a circle is central if its vertex lies at the center of the circle.

   An angle is circumscribed around a circle if its vertex is outside the circle and its sides are tangent to the circle.

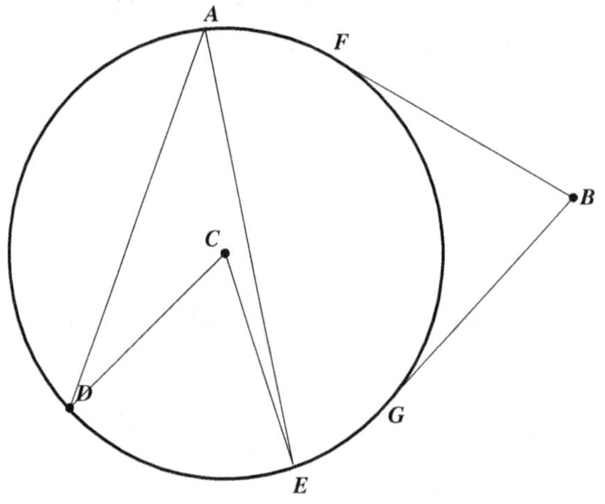

   In circle C above, $\angle DAE$ is inscribed, $\angle DCE$ is central, and $\angle FBG$ is circumscribed because its sides are tangent to the circle (at F and G).

   – What is an arc? How do you measure arcs?

   An arc is a portion of a circle lying between two points on the circle (called "endpoints" of the arc). Because we are on a circle, if you pick two points, you have not uniquely chosen one arc. In fact, unless the two points are endpoints of a diameter, the points divide the circle into a larger arc ("major arc") and a smaller arc ("minor arc"). To avoid confusion, major arcs usually require you to label a point that lies between the two endpoints on the major arc, and then refer to the major arc by those three points. For example, in the picture above, we refer to minor arc DE as $\overset{\frown}{DE}$, the bottom part of the circle between D and E. Minor arcs may only use two letters. If we want the larger, upper part of the circle between D and E, the major arc, then we write $\overset{\frown}{DAE}$, $\overset{\frown}{DFE}$, or $\overset{\frown}{DGE}$. Major arcs require three letters in the notation.

Arcs are measured by a number from 0 to 360, where 360 means the entire circle, 180 means half the circle, 90 means one-quarter of the circle, etc. The measure of $\overset{\frown}{DE}$ is denoted $m\,\overset{\frown}{DE}$. Soon, we will see the connection between this number and angle measure.

– What are some relationships between central, inscribed, and circumscribed angles?

  Many of these angle measurements rely on the concept of "intercepted arcs." An arc is intercepted by an angle if it lies in the interior of the angle. With a circumscribed angle, there are two intercepted arcs, one major and one minor.

  The measure of a central angle is the same as the measure of its intercepted minor arc. A diameter is said to be a central angle of 180 degrees because it intercepts half of the circle. (Which half is technically the intercepted half is not important in this case.) Incidentally, this is usually given as the way to define the measure of a minor arc. The corresponding major arc is then 360 minus the measure of the minor arc.

  The measure of an inscribed angle is one-half of the measure of its intercepted arc.

  The measure of a circumscribed angle is one-half of the difference between its major intercepted arc and its minor intercepted arc.

  Valid proofs of these formulas can be found in any good Geometry textbook, or online.

– What are some other types of angles and how are they related to inscribed, central, and circumscribed angles?

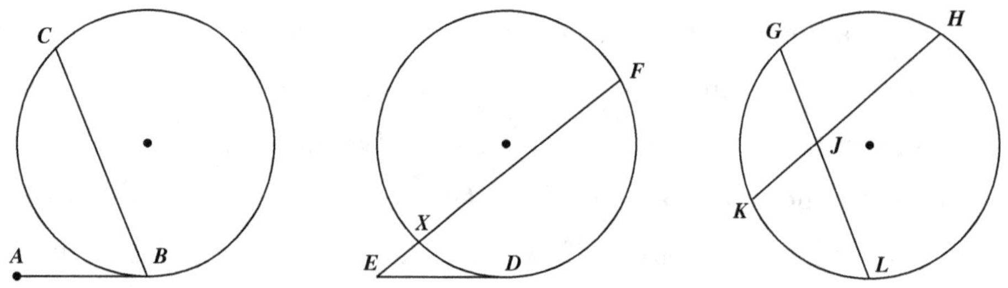

One type of angle is similar to an inscribed angle in that its vertex lies on a given circle, but its sides are different. One of its sides contains a chord of the circle and the other side is tangent to the circle. The concept of an intercepted arc still applies, and this angle's measure is equal to one half the measure of its intercepted arc (just like inscribed angle measure). So in the picture above, $m\angle ABC = \frac{1}{2}m\,\overset{\frown}{BC}$.

Another type of angle is similar to a circumscribed angle in that its vertex lies outside the circle, but its sides are not both tangent. In fact, either both of its sides contain chords of the circle, or one of its sides contains a chord of the circle and the other side is tangent to the circle. In each of these cases, there are two intercepted arcs, and the angle measure is one half the difference of the farther arc and the closer arc (just like circumscribed angle measure). So in the picture above, $m\angle DEF = \frac{1}{2}(m\,\overset{\frown}{DF} - m\,\overset{\frown}{DX})$.

The third type of angle has its vertex inside the circle, but not at the center. Here, two arcs are said to be intercepted, the arc in the interior of the angle, and the arc in the interior of its vertical angle. In this case, the angle measure is one half of the sum of the measures of the two intercepted arcs. So in the picture above, $m\angle GJH = \frac{1}{2}(m\,\overset{\frown}{GH} + m\,\overset{\frown}{KL})$. (Central angles would also satisfy this property - but in the case of a central angle, the two intercepted arcs would be equal. So adding them together and dividing them by two is the same as just looking at the measure of one intercepted arc.)

As before, valid proofs of these formulas can be found in any good Geometry textbook, or online.

– What does it mean for an angle to be inscribed on a diameter?

  An angle is inscribed on a diameter if it intercepts the circle at two endpoints of a diameter, i.e., if its intercepted arc is 180.

– Why are angles inscribed on a diameter right angles?

  Since they are inscribed angles with an intercepted arc measuring 180, they have an angle measure of 90 degrees, which makes them right angles.

- Why is a radius drawn from a point of intersection between a tangent line and a circle necessarily perpendicular to the tangent line?

  There are other ways to prove this, but we will use the angle properties listed above.

  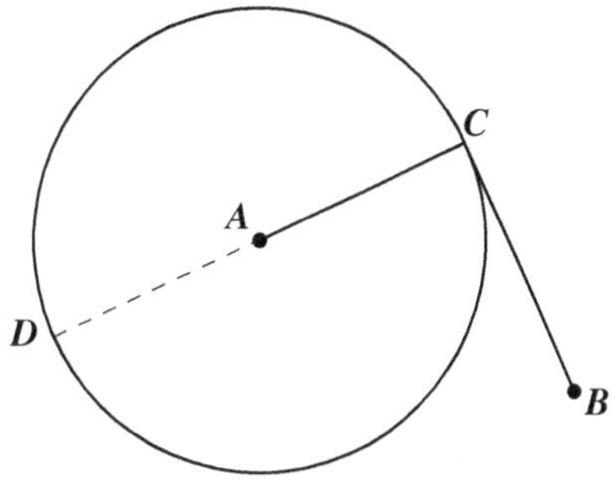

  We need to show that $\angle ACB$ is a right angle if $\overline{BC}$ is tangent to circle A at C. Extend $\overline{CA}$ to form diameter $\overline{CD}$. Then $\angle DCB$ intercepts an arc of measure 180, which means that its measure is half that, or 90 degrees.

  Conversely, suppose that $m\angle ACB = 90$. Then $m\ \overset{\frown}{DC}= 180$, which means that $\overline{DC}$ is a diameter of the circle. We need to show that $\overline{BC}$ is tangent to circle A. So suppose that there is another point of intersection between $\overleftrightarrow{BC}$ and circle A, namely E.

  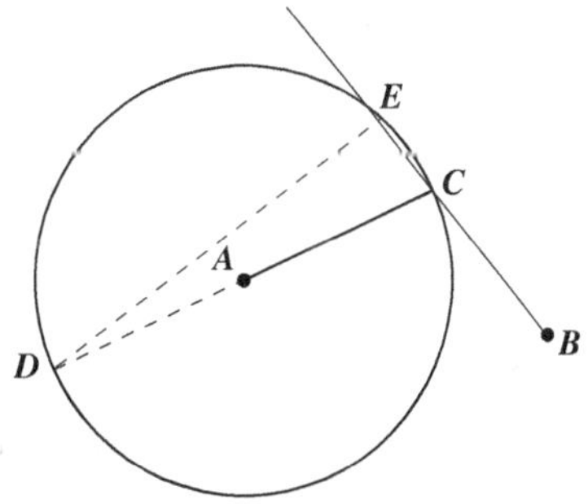

  Consider $\triangle DEC$. We know $m\angle DEC = 90$ because $\angle DEC$ is inscribed in a semicircle. Also, we know $m\angle ECD = 90$ because $\angle ECD$ is supplementary to $\angle ACB$. But this means that $m\angle D = 0$, since the sum of the angle measures in $\triangle DEC$ is 180. Therefore, the picture is not correct. There is no other intersection point between $\overline{BC}$ and circle A. Thus $\overline{BC}$ is tangent to circle A.

3. Construct the inscribed and circumscribed circles of a triangle, and prove properties of angles for a quadrilateral inscribed in a circle.

   - Do we need any theorems before we proceed?

Yes, we do. We need a fact that we haven't proven yet. FACT: The points on an angle bisector are also the points that are equidistant from the sides of the angle. To see this in the picture, draw perpendiculars from M to two sides of the angle, at D and E. We want to show that $\overline{AM}$ bisects $\angle DAE$ if and only if $\overline{MD} \cong \overline{ME}$.

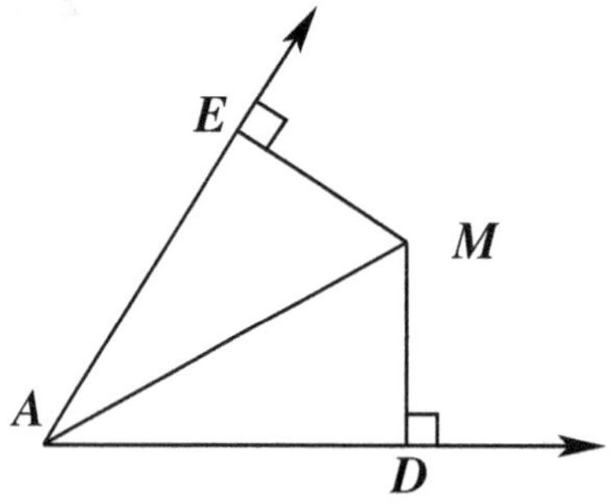

First, suppose $\overline{AM}$ bisects $\angle DAE$. Clearly $\overline{AM} \cong \overline{AM}$. Then by AAS (equivalent to ASA), $\triangle DAM \cong \triangle EAM$. So $\overline{MD} \cong \overline{ME}$. Conversely, suppose that $\overline{MD} \cong \overline{ME}$. Then by HL (Hypotenuse-Leg), $\triangle DAM \cong \triangle EAM$. So $\angle DAM \cong \angle EAM$. Therefore $\overline{AM}$ bisects $\angle DAE$. We have now finished proving the fact, which we will use below.

– Given a triangle, what is the inscribed circle? How do you construct it? Why does the construction work?

  A circle is inscribed in a triangle if the circle is tangent to each of the three sides of the triangle. You can think of it as the largest circle that will fit inside the triangle.

  Constructing any circle involves finding its center and its radius. The center of the inscribed circle (also called the "incircle") is the point where the bisectors of the three angles of the triangle intersect. So construct any two angle bisectors and find their point of intersection (called the "incenter"). To find the radius, we can start from the incenter and construct a line perpendicular to any of the three sides of the triangle. The radius is the distance from the incenter to any of the sides of the triangle.

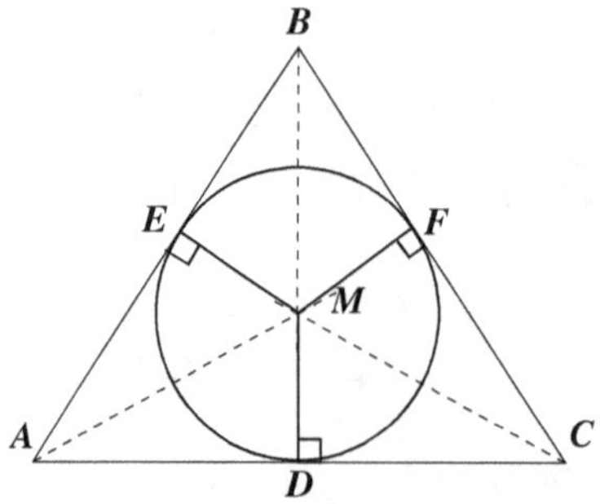

The reason that this construction works hinges on the fact we proved just above. Because point M lies on all three angle bisectors, it is equidistant from all three sides of the triangle. That means that if we draw a circle at M with radius MD, then that circle will be tangent to all three sides of the triangle (at D, E, and F). We used an isosceles triangle in the picture, but it works for any triangle.

- Given a triangle, what is the circumscribed circle? How do you construct it? Why does the construction work?

  A circle is circumscribed around a triangle if it passes through the three vertices of the triangle. You can think of it as the smallest circle that will go around the triangle.

  The circumscribing circle (called the "circumcircle") has its center (called the "circumcenter") at the point where the perpendicular bisectors of the sides of the triangle intersect. Construct perpendicular bisectors of any two sides of the triangle. Their intersection point is the circumcenter. The radius is the distance from the circumcenter to any vertex of the triangle.

  The reason that this construction works is because points on the perpendicular bisector of a segment are equidistant from the endpoints of that segment. This was proved in Congruence Standard #9, above. Hence, if P is the circumcenter of $\triangle ABC$, then $\overline{PA} \cong \overline{PB} \cong \overline{PC}$ by repeatedly invoking this fact. Therefore, if we draw a circle with center P and radius PA, the circle will pass through all three vertices of the triangle, circumscribing it.

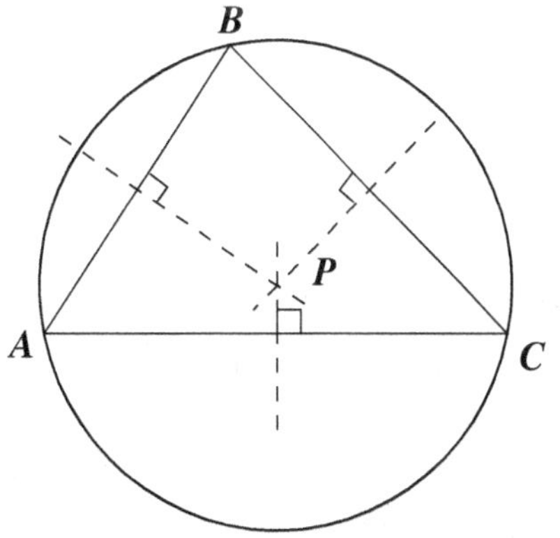

- What are some properties of the angles of a quadrilateral that has been inscribed in a circle?

  Recall that any polygon is said to be "inscribed in a circle" if its vertices lie on that circle. Consider the quadrilateral $ABCD$ inscribed in a circle, below.

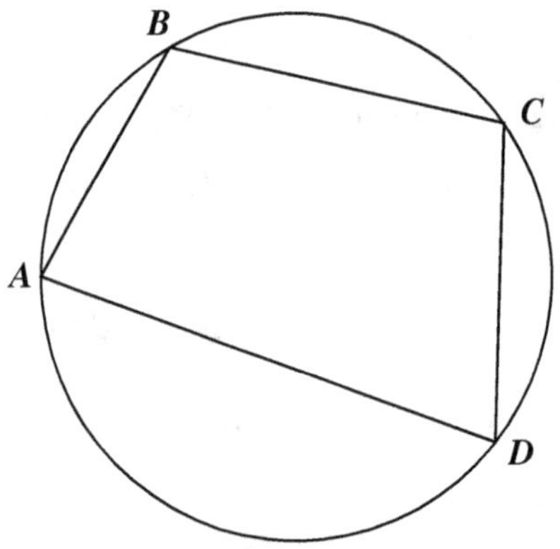

Each of the vertex angles of the quadrilateral is inscribed in the circle. Some consequences are 1) that opposite angles are supplementary, and 2) that the diagonals intersect each other in such a way that the product of the segments of one diagonal equals the product of the segments of the other diagonal.

– How do you prove these properties?

The first property follows from the fact that the arcs intercepted by opposite angles in the quadrilateral are disjoint, but make up the entire circle. For example, $\angle ABC$ intercepts $\overset{\frown}{ADC}$, while $\angle ADC$ intercepts $\overset{\frown}{ABC}$. The whole circle is made up of $\overset{\frown}{ADC}$ and $\overset{\frown}{ABC}$. Thus, $m\angle ABC + m\angle ADC = \frac{360}{2} = 180$ degrees.

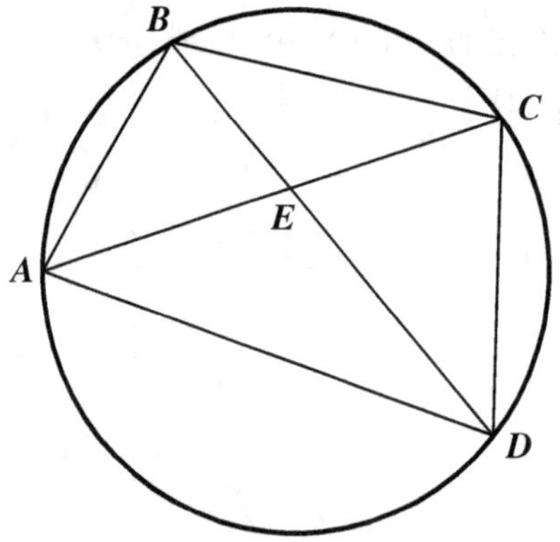

The second property is true of any intersecting chords in a circle and follows from the fact that $\triangle EDA \sim \triangle ECB$ by AA. (We know $\angle AED$ and $\angle BEC$ are vertical; $\angle EAD$ and $\angle EBC$ are both inscribed and share the same intercepted arc, $\overset{\frown}{CD}$.) Hence $\frac{EA}{EB} = \frac{ED}{EC}$, or $(EA)(EC) = (EB)(ED)$.

4. (+) Construct a tangent line from a point outside a given circle to the circle.

– Why are there only two possible tangent segments from a point outside a given circle to the circle? Why are these two segments congruent?

Imagine a circular object (like a can) centered at Q, and a linear object (like a pencil) hinged at a point P outside the circle. Simply spin the object around P until it touches the circle at a point A. Now spin it the other way until it touches the circle at B. These describe the two tangent lines to the circle. The two tangent segments are congruent because they are reflections of one another in $\overleftrightarrow{PQ}$. You can also show that $\triangle PQA \cong \triangle PQB$ by HL ($\overline{QA} \cong \overline{QB}$ because they are radii of circle Q). Either way, $\overline{PA} \cong \overline{PB}$.

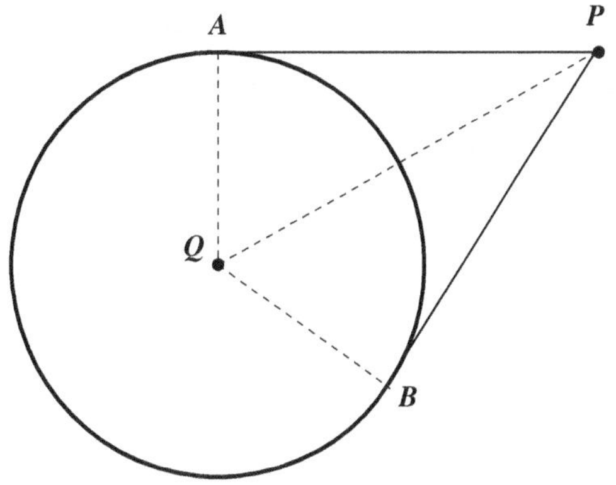

– How do you construct a tangent line from a point outside a given circle to the circle? Why does the construction work?
  As before, suppose P lies outside circle Q. Draw $\overline{PQ}$. Construct its midpoint M. (You can do this by constructing its perpendicular bisector if you wish.) Set your compass to MQ, centered at M, and swing two arcs that intersect the circle, at A and B. Then $\overline{PA}$ and $\overline{PB}$ are tangent segments to circle Q.

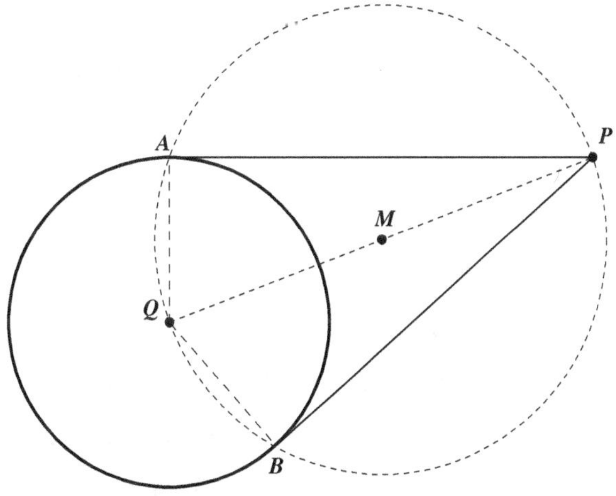

To show that the construction works, we need to show that $\angle PAQ$ is a right angle. But since P, A, and Q all lie on circle M, then $\angle PAQ$ is inscribed in a semicircle, making it a right angle. Since radius $\overline{QA}$ is perpendicular to a tangent line to circle Q at A, it must be that $\overline{PA}$ is the desired tangent segment. The same reasoning shows that $\overline{PB}$ is also tangent to circle Q at B.

- FIND ARC LENGTHS AND AREAS OF SECTORS OF CIRCLES

  5. Derive using similarity the fact that the length of the arc intercepted by an angle is proportional to the radius, and define the radian measure of the angle as the constant of proportionality; derive the formula for the area of a sector.

     – Suppose that we have the same central angle in two different circles. Why is the length of the intercepted arc proportional to the radius of the circle?
     The short answer is because the two arcs are similar. Consider the picture below:

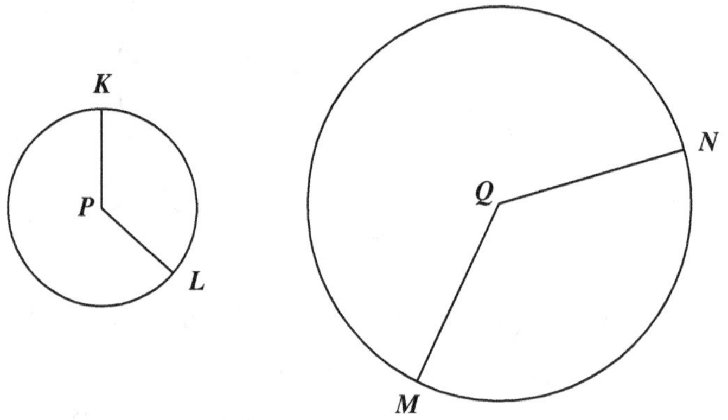

In circle P and in circle Q, we have congruent central angles: $\angle KPL \cong \angle NQM$. Rotate circle P around its center until $\overline{PL}$ is parallel to $\overline{QM}$. Now translate that plane so that P moves to Q. Finally, perform a dilation centered at P with scale factor $k = \frac{QM}{PL}$. Since the dilation preserves the central angle, we have transformed the original $\overset{\frown}{KL}$ of circle P to $\overset{\frown}{NM}$ of circle Q via a similarity transformation. Therefore, the two arcs are similar, which means that their measures are in proportion:

$$\frac{m\,\overset{\frown}{NM}}{m\,\overset{\frown}{KL}} = \frac{QM}{PL} \Longleftrightarrow \frac{m\,\overset{\frown}{NM}}{QM} = \frac{m\,\overset{\frown}{KL}}{PL}.$$

Therefore, in each circle the length of the intercepted arc is proportional to the radius. This is how we define the measure of a central angle in radians. This ratio is constant, regardless of the size of the circle.

     – What does it mean to measure an angle in radians?
     Radians are a more natural way to measure an angle because they are not arbitrary in the way 360 is a somewhat arbitrary number of degrees to use for the total arc measure of a circle.

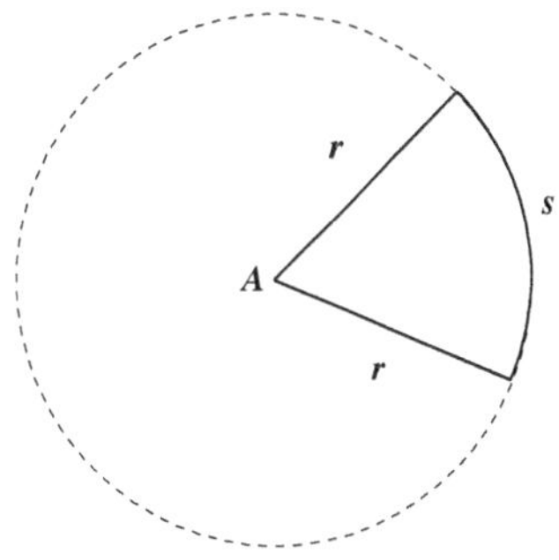

In the picture above, the measure of angle A in radians is defined as a ratio of lengths: $\frac{s}{r}$, where s is the length of the intercepted arc, and r is the radius of the circle. Because of the similarity property discussed above, this angle measure is independent of the size of the circle.

– What is the circumference of a full circle? What is the area of a full circle?
The circumference of a full circle is $2\pi r$, where r is the radius. The area of a full circle is πr^2, where r is the radius. For explanations of where these formulas come from, see the Substandard "Geometric Measurement and Dimension" below. For now, we will simply take these formulas as given.

– How do radians compare with degrees?
Since the circumference of a circle of radius r is $2\pi r$, a full circle angle (360 degrees) equals 2π radians; a half circle (180 degrees) equals π radians; etc. To convert: one radian is $\frac{180}{\pi} \approx 57.3$ degrees, and one degree is $\frac{\pi}{180} \approx 0.0175$ radians.

– What is the formula for the area of a sector of a circle?
If r is the radius of the circle, and if θ is the measure of the sector's central angle in radians, then the area of the sector is $\frac{1}{2}\theta r^2$. Or, if the length of the sector's intercepted arc is s, then the area of the sector is $\frac{1}{2}rs$.

– How do you derive the formula for the area of a sector of a circle?
The formula for the area of a sector of a circle relies on proportional reasoning. If the area of a circle is πr^2, then the area of a semicircle is $\frac{1}{2}\pi r^2$ and the area of a quarter circle is $\frac{1}{4}\pi r^2$, etc. So the fraction in front of πr^2 describes what proportion of the circle is contained in the sector. This is equal to the proportion of the full circular angle that is contained in the central angle of the sector. In other words, if the sector has a central angle of θ radians, then the proportion of

the sector to the full circle is θ to 2π. So if A_{sec} is the area of the sector, then

$$\frac{A_{sec}}{\pi r^2} = \frac{\theta}{2\pi} \Longleftrightarrow A_{sec} = \frac{\theta \pi r^2}{2\pi} = \frac{1}{2}\theta r^2.$$

Since the arc length s equals $r\theta$, then the formula can also be written as $\frac{1}{2}rs$.

EXPRESSING GEOMETRIC PROPERTIES WITH EQUATIONS

- TRANSLATE BETWEEN THE GEOMETRIC DESCRIPTION AND THE EQUATION FOR A CONIC SECTION

1. Derive the equation of a circle of given center and radius using the Pythagorean Theorem; complete the square to find the center and radius of a circle given by an equation.

   – What was the definition of a circle again?

     A circle is the set of all points in the plane that are a given distance (called the radius) from a given point (called the center).

   – How can we use the Pythagorean Theorem to define a Distance Formula?

     The Distance Formula is an important formula in coordinate geometry, but it is really just the Pythagorean Theorem! We will derive it generally now, but use a specific example below.

     Suppose we wish to find the distance between (s, t) and (h, k). We first draw a right triangle with hypotenuse going from (s, t) to (h, k). For the picture, $s > h$ and $t > k$, but for the formula, it doesn't really matter.

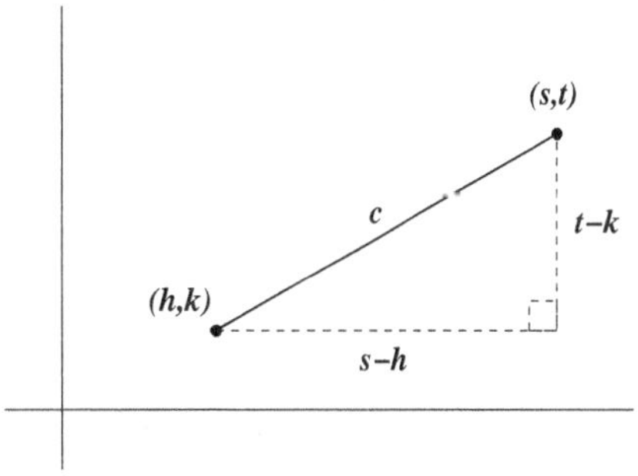

     So, using the Pythagorean Theorem, the hypotenuse c satisfies: $c^2 = a^2 + b^2$.

$$c^2 = (s-h)^2 + (t-k)^2 \Longrightarrow c = \sqrt{(s-h)^2 + (t-k)^2}.$$

     This is the Distance Formula. You can see that it doesn't matter if $s > h$ or $t > k$; their corresponding differences are squared.

   – How can we use the definition to derive the equation of the circle?

     As an example, suppose that the center of the circle is at $(-1, 2)$ and the radius is 5. What is the equation of the circle? We need an equation that is satisfied by those points that are 5 units away from $(-1, 2)$ and only by those points. So,

(x, y) is on the circle if and only if the distance between (x, y) and $(-1, 2)$ is 5. Using the Distance Formula:

$$5 = \sqrt{(x - (-1))^2 + (y - 2)^2} \Longrightarrow (x + 1)^2 + (y - 2)^2 = 25.$$

You can certainly square out the binomials if you wish, but as we will see, this format is much better for circles.

- How can you read off the center and radius from the equation of a circle?
  If the equation of a circle is in its standard form: $(x - h)^2 + (y - k)^2 = r^2$, then this means the circle has radius r and is centered at the point (h, k). For example, if you are asked to describe the points that satisfy $x^2 + (y + 5)^2 = 4$, you could say with certainty that this equation describes a circle of radius 2, centered at the point $(0, -5)$.
- Why do we have to complete the square?
  (For a review of how to complete the square, see the Algebra standard, Seeing Structure in Expressions, #3.)
  Given that the standard form of the equation of the circle is so helpful, we often want to put other equations into the standard form. This may require completing the square. For example, suppose you know that $x^2 + y^2 - 6x + 2y = 0$ describes a circle. What is the center of that circle? What is its radius? We have to complete the square to find out.

$$\begin{aligned} x^2 + y^2 - 6x + 2y &= 0 \\ x^2 - 6x + y^2 + 2y &= 0 \\ x^2 - 6x + \underline{9} + y^2 + 2y + \underline{1} &= 0 + \underline{9} + \underline{1} \\ (x - 3)^2 + (y + 1)^2 &= 10 \end{aligned}$$

  (We underlined the numbers that had to be added to both sides of the equation to complete the square.) So, the circle has a radius of $\sqrt{10}$ and is centered at $(3, -1)$.

2. Derive the equation of a parabola given a focus and directrix.

   - What is the definition of a parabola?

A parabola is the set of all points that are the same distance from a given point (called the focus) as they are from a given line (called the directrix). The distance from a point on the parabola to the directrix is measured perpendicularly to the directrix. For simplicity, we will restrict ourselves only to horizontal or vertical directrices.

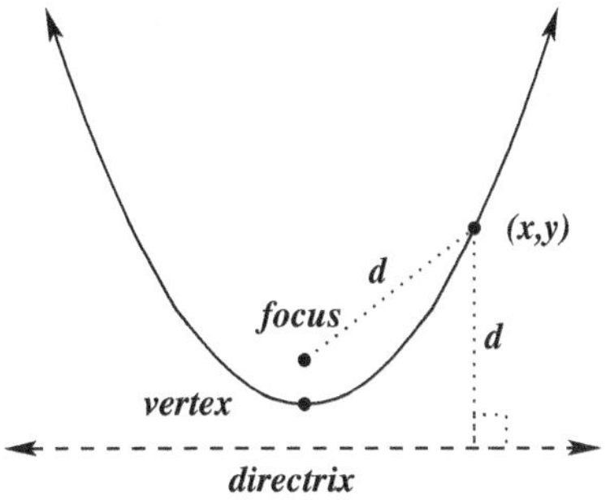

– What is the Vertex Form of the equation of a parabola?

For up or down parabolas with a vertex at (h, k), the equation is of the form:

$$y - k = a(x - h)^2,$$

where the parabola opens upward if $a > 0$, downward if $a < 0$.

For right or left parabolas with a vertex at (h, k), the equation is of the form:

$$x - h = a(y - k)^2,$$

where the parabola opens to the right if $a > 0$, to the left if $a < 0$.

– How can you find the distance from a point to a line?

We determined the Distance Formula above for the distance between two points. Now we need to determine a formula for the distance from a point to a horizontal or vertical line. These formulas can be seen in the following picture.

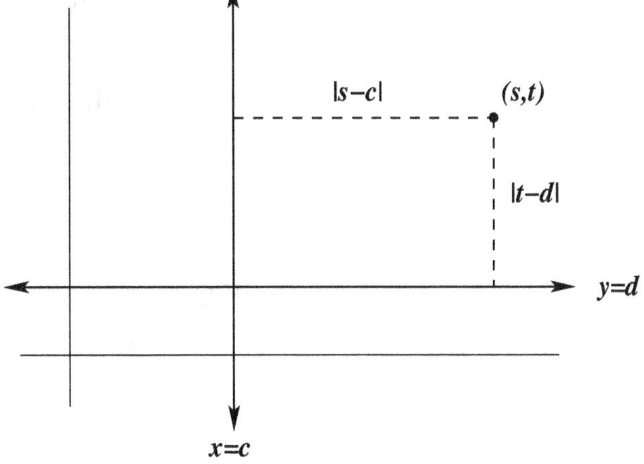

So the distance from (s, t) to the horizontal line $y = d$ is $|t - d|$. Similarly, the distance from (s, t) to the vertical line $x = c$ is $|s - c|$. We need the absolute value in case $s < c$ or $t < d$ (though these cases are not pictured).

- How can you derive the equation of a parabola from its definition?
  We will show a specific example here, but the general formulas are not much harder to obtain. Suppose we have a parabola with a focus at $(1, 4)$ and a directrix of $x = -3$. The parabola is the set of points (x, y) that are as far away from $(1, 4)$ as they are from the line $x = -3$. So

$$\begin{aligned} \text{distance to focus} &= \text{distance to directrix} \\ \sqrt{(x-1)^2 + (y-4)^2} &= |x+3|. \end{aligned}$$

Squaring both sides gives:

$$\begin{aligned} (x-1)^2 + (y-4)^2 &= (x+3)^2 \\ x^2 - 2x + 1 + (y-4)^2 &= x^2 + 6x + 9 \\ (y-4)^2 &= 8x + 8 \\ (y-4)^2 &= 8(x+1) \\ \frac{1}{8}(y-4)^2 &= x + 1 \end{aligned}$$

This is in Vertex Form, with the vertex at $(-1, 4)$ (halfway between the focus and directrix). Notice that this parabola opens to the right.

3. (+) Derive the equations of ellipses and hyperbolas given the foci, using the fact that the sum or difference of distances from the foci is constant.

- What is the definition of an ellipse?
  Given two points called "foci" (plural of "focus"), a point lies on an ellipse if the sum of the distances from the point to the two foci is constant. You can draw an ellipse by attaching a fixed length of string to two foci and then using a pencil to stretch the string taut. The figure you can draw is an ellipse.

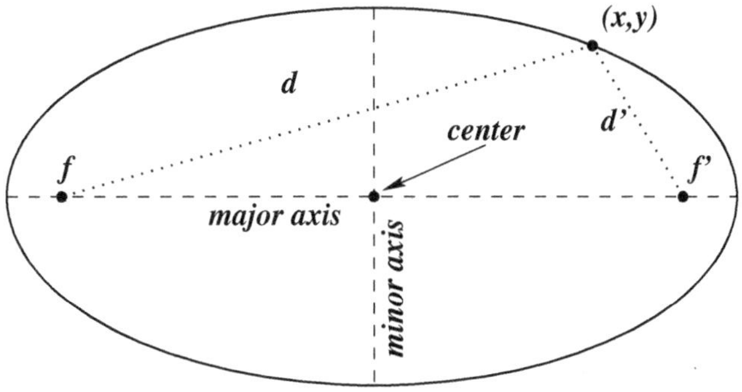

The foci are f and f'. The sum d + d' is constant.

– What is the standard form for the equation of an ellipse?

If an ellipse is centered at (h, k) and has an x-radius of a and a y-radius of b, then its equation is:

$$\frac{(x-h)^2}{a^2} + \frac{(y-k)^2}{b^2} = 1.$$

If $a > b$, as it is in the picture above, then the horizontal distance across the ellipse is called the major axis and the vertical distance is called the minor axis, and vice versa if $b > a$ (not pictured). (If $a = b$, then the ellipse is a circle!) In the picture above, the x-radius is half the length of the major axis and the y-radius is half the length of the minor axis.

– How do you derive the equation of an ellipse from its definition?

Again, we will show a specific example. A good Geometry or Algebra II textbook will have a general derivation. Suppose that the ellipse has foci at $(-4, 0)$ and $(0, 0)$ and that the sum of the distances to the two foci is $2\sqrt{13}$. So, distance from (x, y) to focus 1 plus distance from (x, y) to focus 2 equals $2\sqrt{13}$.

$$\begin{aligned} \sqrt{(x+4)^2 + y^2} + \sqrt{x^2 + y^2} &= 2\sqrt{13} \\ \sqrt{(x+4)^2 + y^2} &= 2\sqrt{13} - \sqrt{x^2 + y^2}. \end{aligned}$$

Squaring both sides gives:

$$\begin{aligned} (x+4)^2 + y^2 &= 52 - 4\sqrt{13(x^2 + y^2)} + x^2 + y^2 \\ 8x - 36 &= -4\sqrt{13(x^2 + y^2)} \\ 2x - 9 &= -\sqrt{13x^2 + 13y^2}. \end{aligned}$$

Again, squaring both sides, we get:

$$\begin{aligned} 4x^2 - 36x + 81 &= 13x^2 + 13y^2 \\ 81 &= 9x^2 + 36x + 13y^2. \end{aligned}$$

Completing the square, we add 36 to each side:

$$\begin{aligned} 81+36 &= 9(x^2+4x+4)+13y^2 \\ 117 &= 9(x+2)^2+13y^2 \\ 1 &= \frac{(x+2)^2}{13}+\frac{y^2}{9}. \end{aligned}$$

This is the standard form for the equation of an ellipse. From it, you can see that its center is at $(-2,0)$ (halfway between the foci) and that its x-radius is $\sqrt{13}$ and its y-radius is $\sqrt{9}=3$.

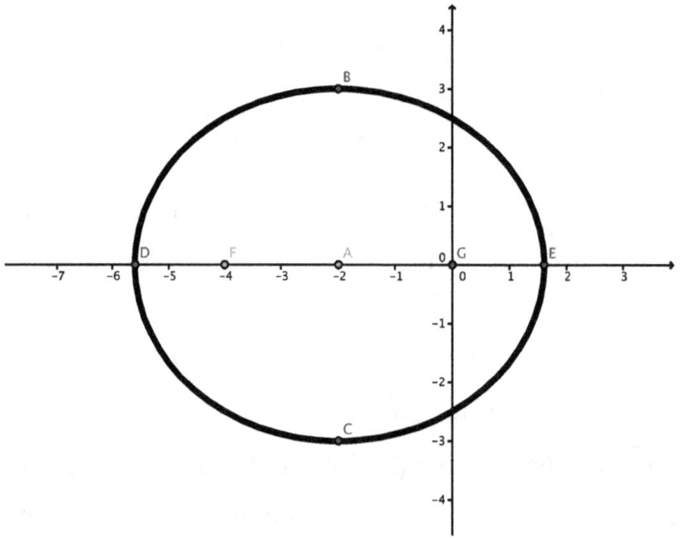

– What is the definition of a hyperbola?

Given two points called foci, a point lies on a hyperbola if the difference of the distances from the point to the two foci is constant.

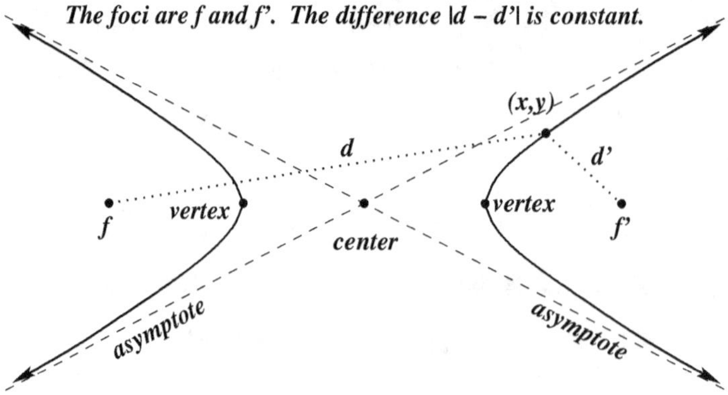

– What are some features of a hyperbola?

A hyperbola has two symmetric branches, and two asymptotes. Each end of each branch approaches one of the asymptotes, getting closer as you move farther from the center. We will assume our hyperbolas are in "standard form," meaning the branches either open up and down, or right and left. In either case, the asymptotes have slopes that are opposites of one another.

- What is the standard form for the equation of a hyperbola?
  For an up-down hyperbola with a center at (h, k), and an x-radius of a, and a y-radius of b, the equation is

$$\frac{(y-k)^2}{b^2} - \frac{(x-h)^2}{a^2} = 1,$$

  in which case the asymptotes have slopes $\pm\frac{b}{a}$ and intersect at the center. The vertices of this hyperbola are at $(h, k+b)$ and $(h, k-b)$. The term x-radius might be misleading here, since the hyperbola doesn't pass to the right or left of the center, but the value is helpful in determining that the slopes of the asymptotes are $\pm\frac{b}{a}$.
  For a left-right hyperbola with a center at (h, k), and an x-radius of a, and a y-radius of b, the equation is

$$\frac{(x-h)^2}{a^2} - \frac{(y-k)^2}{b^2} = 1,$$

  in which case the asymptotes have slopes $\pm\frac{b}{a}$ and intersect at the center. The vertices of this hyperbola are at $(h+a, k)$ and $(h-a, k)$. The term y-radius might be misleading here, since the hyperbola does not pass above or below the center, but the value is helpful in determining that the slopes of the asymptotes are $\pm\frac{b}{a}$.

- How do you derive the equation of a hyperbola from its definition?
  As with the ellipse, the derivation of the hyperbola equation is straightforward, but requires a lot of algebra. Again, we will show a specific example and refer the reader to a good Geometry or Algebra II textbook for a general derivation. Suppose that the hyperbola has foci at $(0, 5)$ and $(0, -1)$ and that the difference of the distances from any point on the hyperbola to the two foci is 4. So, distance from (x, y) to focus 1 minus the distance to focus 2 equals 4.

$$\begin{aligned}
\sqrt{x^2+(y-5)^2} - \sqrt{x^2+(y+1)^2} &= 4 \\
\sqrt{x^2+(y-5)^2} &= 4 + \sqrt{x^2+(y+1)^2}
\end{aligned}$$

Squaring both sides:

$$\begin{aligned} x^2+y^2-10y+25 &= 16+8\sqrt{x^2+(y+1)^2}+x^2+y^2+2y+1 \\ -8\sqrt{x^2+(y+1)^2} &= 12y-8 \\ -2\sqrt{x^2+(y+1)^2} &= 3y-2 \end{aligned}$$

Squaring again:

$$\begin{aligned} 4(x^2+y^2+2y+1) &= 9y^2-12y+4 \\ 4x^2-5y^2+20y &= 0 \end{aligned}$$

So now we complete the square (being very careful with signs):

$$\begin{aligned} 4x^2-5(y^2-4y+4) &= 0-20 \\ 4x^2-5(y-2)^2 &= -20 \\ \frac{4x^2}{-20}-\frac{5(y-2)^2}{-20} &= 1 \\ \frac{(y-2)^2}{4}-\frac{x^2}{5} &= 1. \end{aligned}$$

This tells us the center of the hyperbola is at $(0,2)$ (halfway between the foci), the vertices are at $(0,4)$ and $(0,0)$, and the asymptotes have slopes $\pm\frac{2}{\sqrt{5}}$ and pass through $(0,2)$. See the picture below.

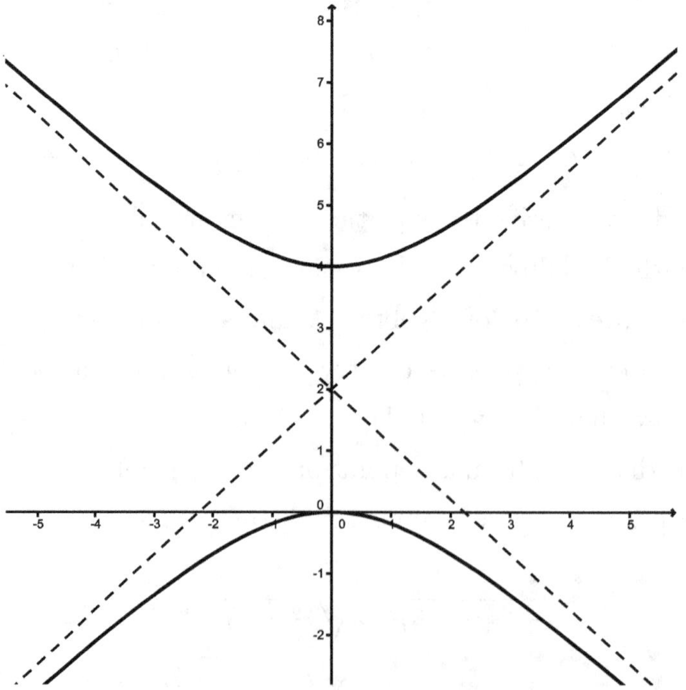

- USE COORDINATES TO PROVE SIMPLE GEOMETRIC THEOREMS ALGEBRAICALLY

4. Use coordinates to prove simple geometric theorems algebraically. *For example, prove or disprove that a figure defined by four given points in the coordinate plane is a rectangle; prove or disprove that the point* $(1, \sqrt{3})$ *lies on the circle centered at the origin and containing the point* $(0, 2)$.

   - What is coordinate geometry? How does it work?

     Coordinate geometry is the application of algebra to geometry. Often, we will place geometric objects on a Cartesian coordinate system and then use algebraic reasoning to discover, justify, show, or even prove a certain relationship.

   - What are some important formulas in coordinate geometry?

     * The Pythagorean Theorem is important. It says that a triangle with side lengths a, b, and c (with c largest) is a right triangle if and only if $a^2+b^2=c^2$.
     * The Distance Formula is also important. We just found above that the distance from (x, y) to (a, b) is $\sqrt{(x-a)^2+(y-b)^2}$.
     * The Midpoint Formula is also important, and will be deduced and generalized below. The midpoint of the line segment from (x, y) to (a, b) is $\left(\frac{x+a}{2}, \frac{y+b}{2}\right)$.
     * The Slope Formula will also play a big role. The slope of the line segment from (x, y) to (a, b) is $\frac{y-b}{x-a}$. If $x = a$, then the line is vertical and its slope is not defined.
     * The Slope-Intercept form of an equation of a line is: $y = mx + b$, where m is the slope of the line, and b is the y-intercept.
     * The Point-Slope form of an equation of a line is: $(y - k) = m(x - h)$, where m is the slope and (h, k) is a point on the line.
     * The Standard form of an equation of a line is: $Ax + By = C$, where A, B, and C are constants. Vertical lines can be written in standard form as $x = h$ for some constant h.

     A word of advice: if your proof uses midpoints, then start with generic points like $(2a, 2b)$ instead of (a, b). You are still being completely arbitrary, but the math will be easier.

   - How can you use coordinates to prove or disprove that a figure defined by four given points in the coordinate plane is a rectangle?

One way is to show that the lines containing the sides that meet at each vertex are perpendicular to each other. This will work, once we verify in #5 below what those perpendicular lines look like in coordinate geometry.

Another way to see that a figure defined by four given points is a rectangle is to check that any set of three points makes a right triangle, using the Pythagorean Theorem. For example, consider the quadrilateral with vertices at $A = (0, 3)$, $B = (1, 1)$, $C = (-3, -1)$, and $D = (-4, 1)$. Is $ABCD$ a rectangle?

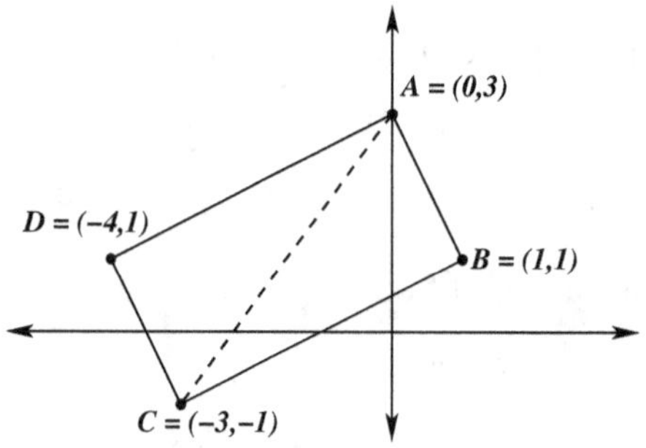

We start by considering $\triangle ABC$. If $ABCD$ is a rectangle, then $\triangle ABC$ should be a right triangle with hypotenuse $\overline{AC}$. We have

$$(AB)^2+(BC)^2 = (0-1)^2+(1-3)^2+(-3-1)^2+(-1-1)^2 = 1+4+16+4 = 25,$$

whereas $(AC)^2 = (-3-0)^2+(-1-3)^2 = 9+16 = 25$. So $\triangle ABC$ is a right triangle, and $\angle B$ is a right angle. Continuing in this way, one can check that $\triangle BCD$, $\triangle CDA$, and $\triangle DAB$ are all right triangles. Therefore $ABCD$ is a rectangle.

– How can you use coordinates to prove or disprove that the point $(1, \sqrt{3})$ lies on the circle centered at the origin and containing the point $(0, 2)$?

One way to determine this is to figure out the radius of the circle (since we already know the center is at the origin). We are told that $(0, 2)$ lies on the circle. Using the Distance Formula from $(0, 0)$ to $(0, 2)$ (or just eyeballing it on a graph), we find that the radius of the circle is 2. So the question becomes, "Is the point $(1, \sqrt{3})$ two units from the origin?" We can answer this using the distance formula again:

$$\sqrt{(1-0)^2+(\sqrt{3}-0)^2} = \sqrt{1+3} = \sqrt{4} = 2.$$

So, yes, the point $(1, \sqrt{3})$ lies on the circle centered at the origin and containing the point $(0, 2)$.

Another approach, which is really the same thing, is to figure out the equation of the circle centered at the origin, with radius 2: $x^2 + y^2 = 4$. Now we figure out if

$(1, \sqrt{3})$ satisfies this equation. If so, then $(1, \sqrt{3})$ lies on the circle; if not, then it doesn't.

$$1^2 + (\sqrt{3})^2 = 1 + 3 = 4.$$

So again, we see that $(1, \sqrt{3})$ lies on the circle. Notice that the algebra we did was essentially the same in each case.

- What are some other proofs you can do?

  As another example, we will show that the diagonals of a rectangle are congruent. To prove this, we first need to place an arbitrary rectangle on a coordinate system. Since rotating and translating the rectangle will not affect the length of its diagonals, we can place the rectangle so that its vertices are at $(0, 0)$, $(a, 0)$, (a, b), and $(0, b)$, where a and b are positive real numbers.

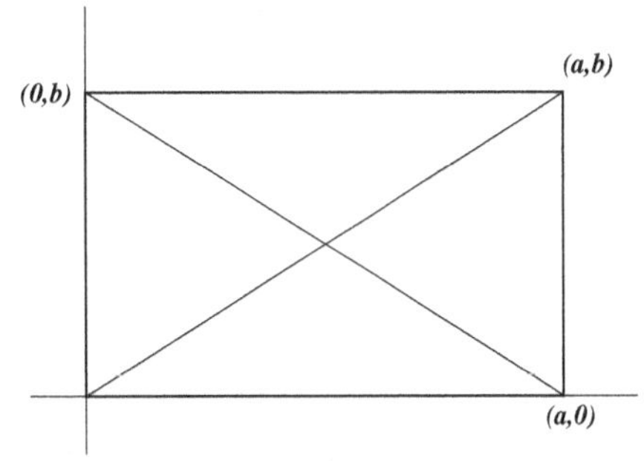

  Using the distance formula, the length of the diagonal from $(0, 0)$ to (a, b) is found to equal $\sqrt{(a-0)^2 + (b-0)^2} = \sqrt{a^2+b^2}$. The other diagonal has length $\sqrt{(a-0)^2 + (0-b)^2} = \sqrt{a^2+b^2}$, which is equal to the length of the first diagonal. Therefore, the diagonals of a rectangle are congruent.

5. Prove the slope criteria for parallel and perpendicular lines and use them to solve geometric problems (e.g., find the equation of a line parallel or perpendicular to a given line that passes through a given point).

   - What are the slope criteria for parallel lines? ...for perpendicular lines?

     Two distinct lines are parallel if 1) they are both vertical, or 2) they have the same slope. (Recall that slope is not defined for vertical lines.) Two lines are perpendicular if 1) one is vertical and one is horizontal, or 2) the product of their slopes is -1. Another way to say this is that their slopes are negative reciprocals of each other.

   - How do you prove the slope criteria for parallel lines?

     Recall that in geometry, lines are parallel if they do not intersect. Algebraically, a point of intersection is a point that satisfies the equations of both lines. So how do we know that two lines with the same slope do not intersect?

Let's start with lines that do not have a defined slope. If you have two distinct vertical lines, like $x = a$ and $x = b$ (with $a \neq b$), then there are no simultaneous solutions to both equations. If $x = a$, then x cannot also equal b simultaneously. So, distinct vertical lines are parallel to each other.

Also, if you have a vertical line, like $x = a$ and a line with a defined slope, line $y = mx + b$, then these lines DO intersect. The common solution to these equations is the point $(a, ma + b)$. Therefore a vertical line CANNOT be parallel to a line with a defined slope.

Now, suppose that two distinct lines have the same slope: $y = mx + b$ and $y = mx + c$ (with $b \neq c$). We know that these equations cannot have a common solution. For if $y = mx+b = mx+c$, then $b = c$, which really means that the lines are the same. So if two distinct lines have the same slope, they do not intersect, and are therefore parallel.

Finally, suppose that we have two distinct lines with different slopes: $y = mx + b$ and $y = nx + c$, with $m \neq n$. It turns out that these lines DO intersect, and so these lines are NOT parallel. To find the intersection point from this general formula requires solving two equations simultaneously. So,

$$\begin{aligned} y = mx + b &= nx + c \\ mx - nx &= c - b \\ (m - n)x &= c - b \\ x &= \frac{c - b}{m - n}. \end{aligned}$$

Since $m \neq n$, the denominator cannot be zero, and thus x is defined. From here, we can substitute (and simplify by getting a common denominator) to find that

$$y = m\left(\frac{c - b}{m - n}\right) + b = \frac{cm - bm + bm - bn}{m - n} = \frac{cm - bn}{m - n}.$$

Therefore, by exhausting all the possibilities, we have shown that two distinct lines are parallel if and only if 1) they are both vertical, or 2) they have the same slope.

– How do you prove the slope criteria for perpendicular lines?

If our two lines do not intersect, then they are parallel and definitely not perpendicular. So suppose we have two lines that intersect at a point. Using a translation (which doesn't change slopes), we can move the lines so that the point of intersection is at the origin. This will make the algebra easier. Let's start with two lines that have defined slopes. So we have two lines $y = mx$ and $y = nx$, with $m \neq n$, and we want to find out exactly when they are perpendicular. There are a number of ways to proceed, but we'll use the Pythagorean Theorem here. If we pick a point R on one line and S on the other, and call the origin P, then we have to show that $\triangle RPS$ is a right triangle with a right angle at P.

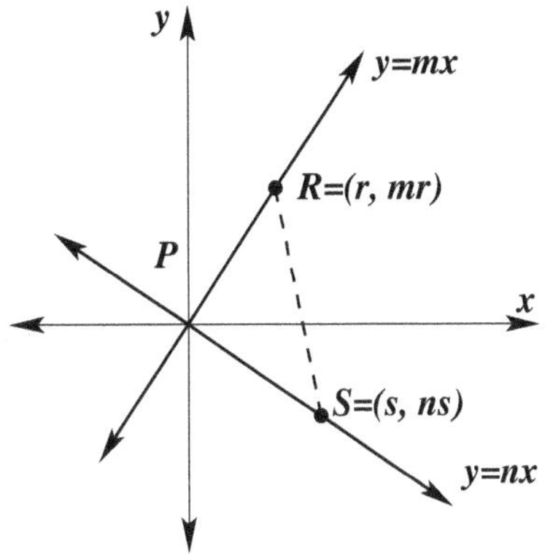

In other words, we need $(RS)^2 = (RP)^2 + (PS)^2$. So let's pick $R = (r, mr)$ on line 1 ($r \neq 0$) and $S = (s, ns)$ on line 2 ($s \neq 0$), and then compare $(RP)^2 + (PS)^2$ to $(RS)^2$. We have

$$\begin{aligned}(RP)^2 + (PS)^2 &= (r-0)^2 + (mr-0)^2 + (s-0)^2 + (ns-0)^2 \\ &= r^2 + m^2r^2 + s^2 + n^2s^2,\end{aligned}$$

whereas:

$$\begin{aligned}(RS)^2 &= (r-s)^2 + (mr-ns)^2 = r^2 - 2rs + s^2 + m^2r^2 - 2mnrs + n^2s^2 \\ &= r^2 + m^2r^2 + s^2 + n^2s^2 - 2rs(1+mn).\end{aligned}$$

Since $rs \neq 0$, these two expressions are equal if and only if $(1 + mn) = 0$. So the lines are perpendicular if and only if $mn = -1$.

Now suppose one of the lines is vertical. We can still use the Pythagorean Theorem. If line 1 is now $x = 0$ (recall that we translated the intersection point to the origin), and the other line is $y = mx$, then what would happen? We can still pick $R = (r, mr)$ on the non-vertical line (with $r \neq 0$), and pick $T = (0, t)$ on the vertical line (with $t \neq 0$).

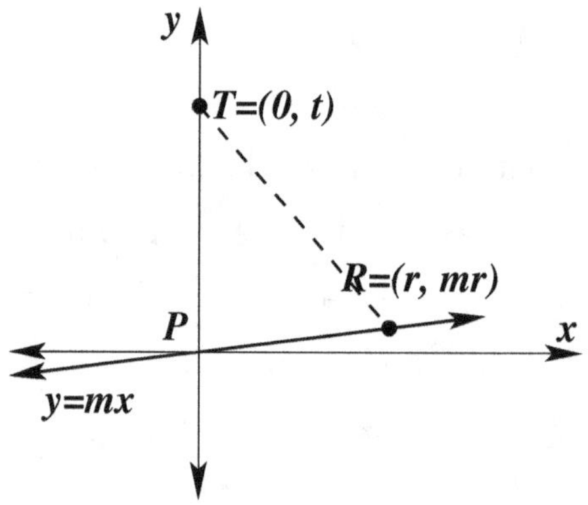

We want to know when $\triangle TPR$ is a right triangle with a right angle at P. So we will compare $(RP)^2 + (TP)^2$ with $(RT)^2$.

$$\begin{aligned}(RP)^2 + (TP)^2 &= (r-0)^2 + (mr-0)^2 + (0-0)^2 + (t-0)^2 \\ &= r^2 + m^2r^2 + t^2,\end{aligned}$$

whereas:

$$\begin{aligned}(RT)^2 &= (r-0)^2 + (mr-t)^2 = r^2 + m^2r^2 - 2mrt + t^2 \\ &= r^2 + m^2r^2 + t^2 - 2mrt.\end{aligned}$$

Since $rt \neq 0$, these two expressions are equal if and only if $m = 0$. In other words, the only line that is perpendicular to a vertical line is a line with slope equal to 0, i.e. a horizontal line.

Therefore, by exhausting all the possibilities, we have shown that two lines are perpendicular if 1) one is vertical and one is horizontal, or 2) the product of their slopes is -1.

- How do you find the equation of a line parallel to a given line that passes through a given point?

  We'll do an example. Suppose you need to find a line that is parallel to the line through $(1, -3)$ and $(4, 2)$, but passes through the point $(1, 5)$.

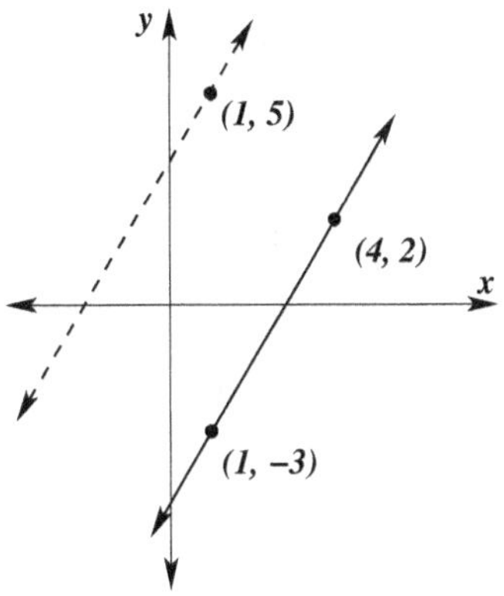

First, let's find the slope of the line we need. Since we need a line parallel to the one through $(1, -3)$ and $(4, 2)$, the slope we want is the same as this slope. Using the Slope Formula, we find that the slope is $\frac{2+3}{4-1} = \frac{5}{3}$. So we need a line with slope $\frac{5}{3}$ and passing through the point $(1, 5)$. Using the Point-Slope form of the line, we get

$$y - 5 = \frac{5}{3}(x - 1).$$

– How do you find the equation of a line perpendicular to a given line that passes through a given point?

We'll do a similar example as above. Suppose you need to find a line that is perpendicular to the line through $(1, -3)$ and $(4, 2)$, but passes through the point $(1, 5)$.

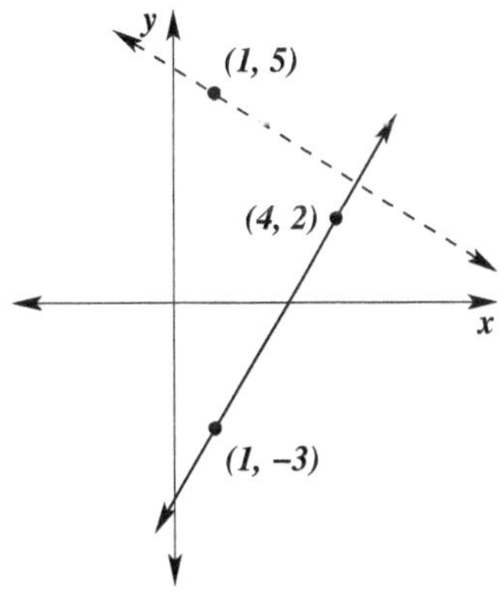

The slope through $(1, -3)$ and $(4, 2)$ is $\frac{2+3}{4-1} = \frac{5}{3}$. Since we want a line perpendicular to this one, the slope we want is $-\frac{3}{5}$. So we need a line with slope $-\frac{3}{5}$ and passing through the point $(1, 5)$. Using the Point-Slope form of the line, we get

$$y - 5 = -\frac{3}{5}(x - 1).$$

6. Find the point on a directed line segment between two given points that partitions the segment in a given ratio.

– What was the Midpoint Formula again?

The midpoint of the line segment from (x, y) to (a, b) is $\left(\frac{x+a}{2}, \frac{y+b}{2}\right)$.

– How do you find the point on a directed line segment between two given points that partitions the segment in a given ratio?

The Midpoint Formula finds a point that divides a segment into two equal parts. So if you are looking to divide a segment into two parts in a 1:1 ratio, then the Midpoint Formula will work. However, this question asks us to generalize that formula to other ratios. We'll write down the formula first and then explain why it works. Notice the following (perhaps unusual) way of writing the Midpoint Formula:

$$\left(\frac{x+a}{2}, \frac{y+b}{2}\right) = \left(\frac{1x+1a}{1+1}, \frac{1y+1b}{1+1}\right).$$

If you want to divide the line segment from (x, y) to (a, b) in a ratio that equals $q : p$, then use

$$\left(\frac{px+qa}{p+q}, \frac{py+qb}{p+q}\right).$$

More precisely, if $S = (x, y)$, $T = (a, b)$, then the point $D = \left(\frac{px+qa}{p+q}, \frac{py+qb}{p+q}\right)$ divides $\overline{ST}$ so that $\frac{SD}{DT} = \frac{q}{p}$.

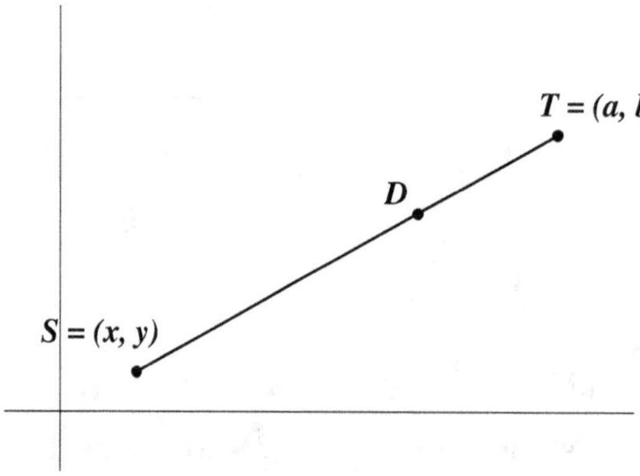

– Why does it work?

We will prove it using the distance formula, but a more advanced way of looking at the picture would require the use of vectors. (See the relevant Common Core State Standards.) We will calculate SD and DT directly. After writing down the

Distance Formula, we will get common denominators in each set of parentheses.

$$\begin{aligned}
SD &= \sqrt{\left(\frac{px+qa}{p+q}-x\right)^2+\left(\frac{py+qb}{p+q}-y\right)^2} \\
&= \sqrt{\left(\frac{px+qa-px-qx}{p+q}\right)^2+\left(\frac{py+qb-py-qy}{p+q}\right)^2} \\
&= \sqrt{\left(\frac{q(a-x)}{p+q}\right)^2+\left(\frac{q(b-y)}{p+q}\right)^2} \\
&= \sqrt{\frac{q^2(a-x)^2}{(p+q)^2}+\frac{q^2(b-y)^2}{(p+q)^2}} = \sqrt{\frac{q^2}{(p+q)^2}\left((a-x)^2+(b-y)^2\right)} \\
&= \frac{q}{p+q}\sqrt{(a-x)^2+(b-y)^2}.
\end{aligned}$$

A similar calculation leads to

$$DT = \frac{p}{p+q}\sqrt{(a-x)^2+(b-y)^2}.$$

Thus

$$\frac{SD}{DT} = \frac{\frac{q}{p+q}\sqrt{(a-x)^2+(b-y)^2}}{\frac{p}{p+q}\sqrt{(a-x)^2+(b-y)^2}} = \frac{q}{p}.$$

7. Use coordinates to compute perimeters of polygons and areas of triangles and rectangles, e.g., using the distance formula.**

   – How do you use coordinates to compute the perimeter of a polygon?

   Since "perimeter" just means the total length of all the sides, you can use the Distance Formula to add up the lengths of the sides of the polygon to get its perimeter.

   As an example, consider the triangle with vertices at the points $A = (1, 1)$, $B = (-2, 5)$, and $C = (0, 7)$. To find its perimeter, we need to find $AB + BC + CA$. Using the Distance Formula, we find:

$$\begin{aligned}
AB &= \sqrt{(-2-1)^2+(5-1)^2} = \sqrt{9+16} = 5; \\
BC &= \sqrt{(0+2)^2+(7-5)^2} = \sqrt{4+4} = \sqrt{8}; \\
CA &= \sqrt{(1-0)^2+(1-7)^2} = \sqrt{1+36} = \sqrt{37}.
\end{aligned}$$

   So the perimeter of $\triangle ABC$ is $5+\sqrt{8}+\sqrt{37} \approx 13.91$.

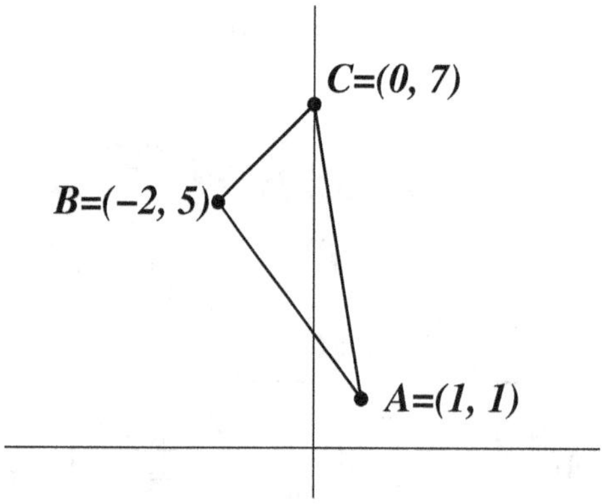

– How do you use coordinates to compute the area of a triangle?

There are several formulas for the area of a triangle, most notably $\frac{1}{2}bh$, where b is the base and h is the height. We can use any of the sides as the base, but determining the height requires some perpendicular lines, unless you are fortunate enough to have one side either horizontal or vertical.

As a general example, consider $\triangle ABC$ given just above, with $A = (1, 1)$, $B = (-2, 5)$, and $C = (0, 7)$. To find the area of $\triangle ABC$, we need to choose a base. Since AB is an integer (5), let's choose AB as the base.

To find the height, we need to find the perpendicular distance from C to $\overleftrightarrow{AB}$. We can do this using the tools above. (Here again, vectors could make the problem a little easier.)

First, let's find the equation of $\overleftrightarrow{AB}$. The slope is $\frac{5-1}{-2-1} = -\frac{4}{3}$. Using point A and starting from the Point-Slope form of the equation of $\overleftrightarrow{AB}$, we get

$$(y-1) = -\frac{4}{3}(x-1) \Longleftrightarrow y = -\frac{4}{3}x + \frac{4}{3} + 1 = -\frac{4}{3}x + \frac{7}{3}.$$

Now let's find the line through C perpendicular to $\overleftrightarrow{AB}$, that is, with a slope of $\frac{3}{4}$, the negative reciprocal of $-\frac{4}{3}$. Using the Point-Slope equation, we get $y - 7 = \frac{3}{4}(x - 0)$, which means $y = \frac{3}{4}x + 7$.

Next we need to find D, the foot of the altitude of $\triangle ABC$ through C. This will be the intersection point of the two lines we just found. So, setting the y components equal:

$$-\frac{4}{3}x + \frac{7}{3} = \frac{3}{4}x + 7.$$

Multiplying by 12 to clear denominators, we obtain:

$$\begin{aligned} -16x + 28 &= 9x + 84 \\ -56 &= 25x \\ x &= -\frac{56}{25}. \end{aligned}$$

From here we can determine that $y = \frac{3}{4}(-\frac{56}{25}) + 7 = -\frac{168}{100} + \frac{700}{100} = \frac{532}{100} = \frac{133}{25}$. So $D = \left(-\frac{56}{25}, \frac{133}{25}\right)$.

Finally, we can compute the height CD with respect to the base $\overline{AB}$. We have:

$$\begin{aligned} CD &= \sqrt{\left(-\frac{56}{25} - 0\right)^2 + \left(\frac{133}{25} - 7\right)^2} \\ &= \sqrt{\left(-\frac{56}{25}\right)^2 + \left(-\frac{42}{25}\right)^2} \\ &= \sqrt{\frac{(56)^2 + (42)^2}{(25)^2}} = \sqrt{\frac{3136 + 1764}{625}} \\ &= \sqrt{\frac{4900}{625}} = \frac{70}{25} = \frac{14}{5}. \end{aligned}$$

Therefore, the area of $\triangle ABC$ is $\frac{1}{2}(AB)(CD) = \frac{1}{2}(5)(\frac{14}{5}) = \frac{14}{2} = 7$. (Whew!)

As a side note, we could have also used Heron's formula for the area of a triangle in terms of its three side lengths, but it is more cumbersome. If the sides have lengths a, b, and c, then Heron's formula says that the area is $\sqrt{s(s-a)(s-b)(s-c)}$, where s is half of the perimeter; that is, $s = \frac{a+b+c}{2}$. Given our perimeter, this path looks rather daunting, although it should also give us an area of 7.

– There's got to be a better way to find the area of a triangle, given the coordinates of its vertices. What is it?

  You could also create triangles that have horizontal or vertical sides within a bounding rectangle. For instance, using the example above, we could have drawn a rectangle around $\triangle ABC$.

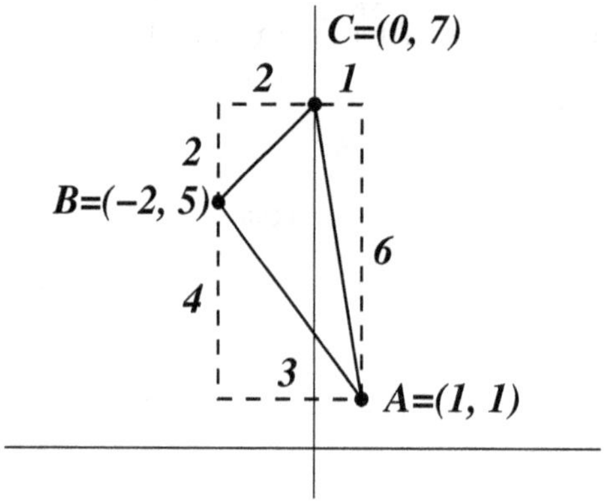

Then, the solution is more straightforward. The entire area of the rectangle is $(6)(3) = 18$. Now we subtract the areas of the three right triangles with dashed legs. Their total areas are:

$$\frac{1}{2}(4)(3) + \frac{1}{2}(2)(2) + \frac{1}{2}(6)(1) = 6 + 2 + 3 = 11.$$

So the area of $\triangle ABC$ must be $18 - 11 = 7$.

– How do you use coordinates to compute the area of a rectangle?

Since the area of a rectangle is the product of its base and its height, we simply need to find two consecutive side lengths and multiply them together. As an example, consider the rectangle $ABCD$, with vertices at $A = (0, 3)$, $B = (1, 1)$, $C = (-3, -1)$, and $D = (-4, 1)$. Its area will be $(AB)(BC)$, or

$$\sqrt{(1-0)^2 + (1-3)^2}\sqrt{(-3-1)^2 + (-1-1)^2} = \sqrt{1+4}\sqrt{16+4} = \sqrt{100} = 10.$$

GEOMETRIC MEASUREMENT AND DIMENSION

- EXPLAIN VOLUME FORMULAS AND USE THEM TO SOLVE PROBLEMS

1. Give an informal argument for the formulas for the circumference of a circle, area of a circle, volume of a cylinder, pyramid, and cone. *Use dissection arguments, Cavalieri's principle, and informal limit arguments.*

   – What is the formula for the circumference of a circle? Why?

   The circumference of a circle is equal to π times the diameter of the circle. Alternatively, the circumference is equal to $2\pi r$, where r is the radius of the circle. We saw earlier that all circles are similar, and that the length of the intercepted arc of a central angle is therefore proportional to the radius of the circle, and thus is proportional to the diameter of the circle as well. So in each circle, the ratio of the circumference to the diameter must be a constant. The name of that constant of proportionality is π. That's how π is defined in Euclidean geometry. So, that's why $C = \pi d = 2\pi r$ (where d is the diameter of the circle).

   – What is the formula for the area of a circle? Why?

   The area of a circle is πr^2, if r is the circle's radius.

   To see why this is the correct formula, we first consider how to use triangles to find the area of a regular polygon.

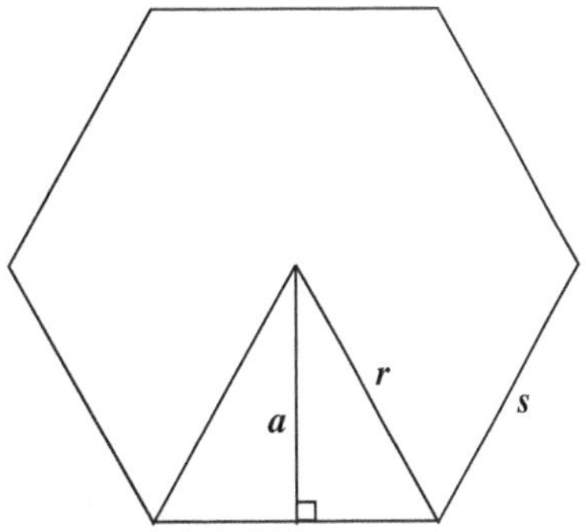

   In the diagram, we are looking at a hexagon, but you can imagine the polygon to have any number of sides. Let n be the number of sides of the polygon. The length a is called the *apothem.* The side length s is just the perimeter P of the polygon divided by n: $s = P/n$. Since the polygon is regular, drawing the radii will divide the polygon into n congruent triangles, each with base s and height a.

So the total area of the polygon is: $A = n(\frac{1}{2}sa) = \frac{1}{2}a(ns) = \frac{1}{2}aP$. Another way to see this formula is:

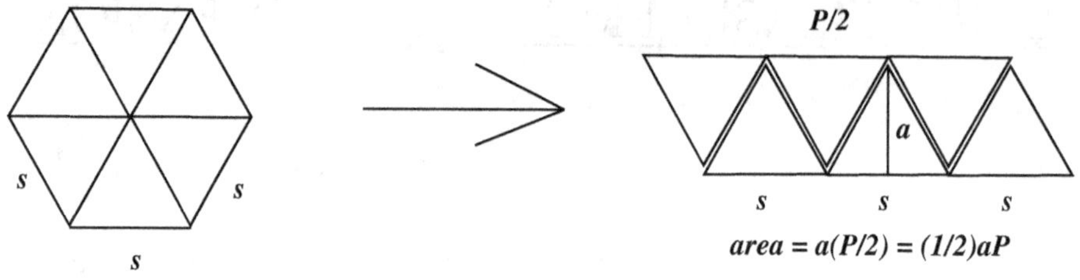

Now, to move towards a circle, think of a regular polygon having more and more sides. As the number of sides increases, the polygon looks more and more like a circle. So, if you think about what happens, you can determine that the apothem approaches the radius ($a \to r$) and the perimeter approaches the circumference ($P \to 2\pi r$) as the number of sections increases (as $n \to \infty$). Therefore,

$$A = \frac{1}{2}aP \to \frac{1}{2}(r)(2\pi r) = \pi r^2.$$

- What are the volume and surface area of a cube?
  If a cube has side length s, then its volume is s^3. In many Geometry textbooks, this is the definition of "volume" and other volume formulas are deduced from this definition.
- What are the other solids and what are their volume formulas?
  The other solids (roughly in order of increasing complexity) are:
  * rectangular prism - congruent rectangle bases lying in parallel planes (one directly aligned with the other) with corresponding vertices joined, making four rectangular lateral faces
  * prism - congruent polygon bases lying on parallel planes with corresponding vertices joined, making parallelogram lateral faces (PRISMS CAN BE SLANTED!)
  * pyramid - polygon base with each vertex joined to a point (called "the" vertex) on a different plane, making triangular lateral faces (PYRAMIDS CAN BE SLANTED!)
  * cylinder - congruent circular bases lying on parallel planes, joined by one curved lateral face (CYLINDERS CAN BE SLANTED!)
  * cone - a circular base joined to a vertex on a different plane, making one curved lateral face (CONES CAN BE SLANTED!)

Solid	Volume
prism	Bh, where B is area of base
pyramid	$\frac{1}{3}Bh$
cylinder	$\pi r^2 h$
cone	$\frac{1}{3}\pi r^2 h$

– How are these volume formulas obtained?

You can obtain all of the volume formulas using calculus. However, you can also find convincing geometric justifications as well. Euclid finds the volumes of the Platonic solids in Book XIII of the *Elements.*

The volumes of a prism and a cylinder are of the form Bh, where B is the area of the base. This is plausible because each slice perpendicular to the height has exactly the same area.

To find the volume of a pyramid, you can subdivide a cube into 3 congruent pyramids of the same base area and height as the original cube, which explains why $\frac{1}{3}$ shows up in the volume of a pyramid.

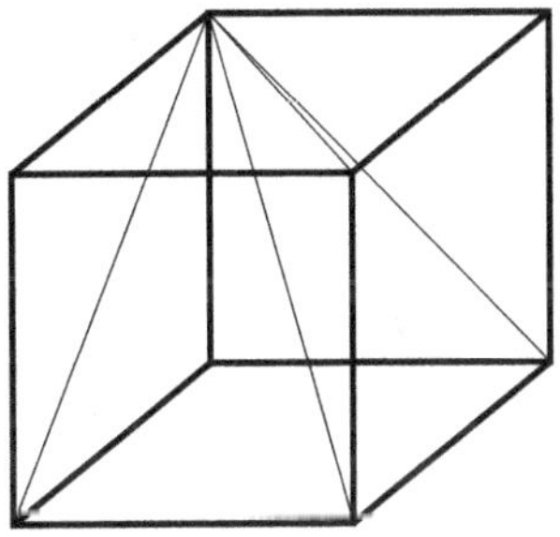

The relationship of a cone to a cylinder is analogous to the relationship of a pyramid to a prism of the same base and height. So, the volume of a cone is one-third that of the cylinder of the same base and height.

2. (+) Give an informal argument using Cavalieri's principle for the formulas for the volume of a sphere and other solid figures.

– What is the formula for the volume of a sphere? Why?

A sphere is the set of all points in space which lie a certain distance r (called the radius) from a given point (called the center). The volume of a sphere is $\frac{4}{3}\pi r^3$.

Archimedes found the volume of a sphere by showing that the ratio of the volume of a sphere to the volume of its circumscribing cylinder is 2:3. He was so proud of this result that he reportedly wanted it put on his tombstone.

Here's how it works. If you take a double cone, a sphere, and a cylinder, all of the same radius r and height $2r$, and if you take circular slices through all three of them at the same height, then the area of the cylinder slice is the sum of the areas of the double cone slice and the sphere slice.

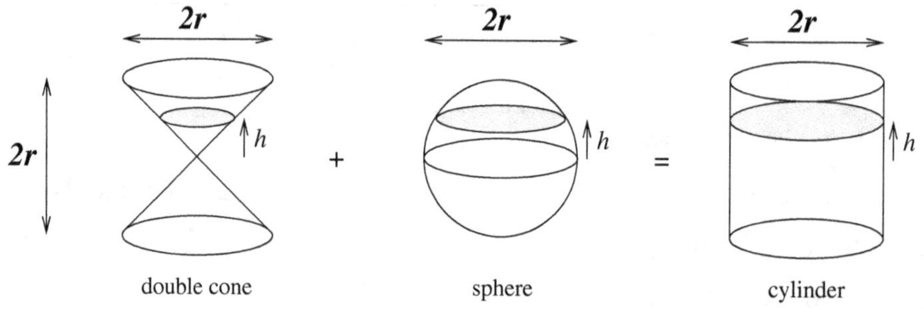

To see this, assume that the double cone, sphere, and cylinder are centered at a height of zero. Hence for the cone, if the slice is at height h, then the radius of the slice is h. For the sphere, however, the radius of the slice is $\sqrt{r^2 - h^2}$, since the side view of the sphere is a circle, also of radius r. So the sum of the areas of these slices is

$$\pi h^2 + \pi(\sqrt{r^2 - h^2})^2 = \pi h^2 + \pi(r^2 - h^2) = \pi r^2,$$

which is precisely the area of the cylinder slice at height h.
So, by Cavalieri's Principle[4], the volume of the double cone plus the volume of the sphere equals the volume of the cylinder. In other words:

$$2\left(\frac{1}{3}\pi r^2 r\right) + V_{sph} = (\pi r^2)(2r).$$

From this, one obtains $V_{sph} = 2\pi r^3 - \frac{2}{3}\pi r^3 = \frac{4}{3}\pi r^3$.

- What is the formula for the volume of a slanted figure, like a cylinder, a cone, etc.? Why?
  Suppose you have a slanted cylinder of base area B and height h (where the height is measured perpendicularly to the base). Next to this slanted cylinder, imagine placing a right cylinder with the same base area B and the same height h. Then imagine a horizontal slice intersecting both cylinders. (The picture below also shows a slanted pyramid and a right pyramid of the same height.)

[4]Cavalieri's Principle says that if you take two solids and if parallel planes through those solids always intersect both solids so that the areas of intersection are equal, then the volumes of the solids must be equal.

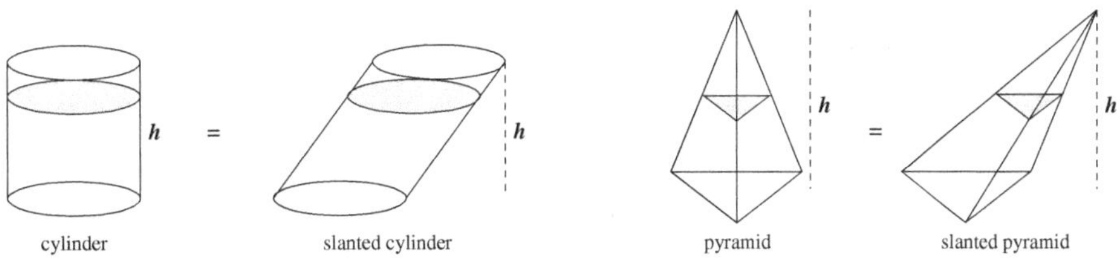

The area of the slice of the right cylinder equals the area of the slice of the slanted cylinder. So, using Cavalieri's Principle, these two cylinders have the same volume. A similar argument shows that the volume of any slanted object is the same as the volume of its corresponding right object. So the volume formulas listed above also work for slanted prisms, pyramids, cylinders, and cones, provided that the height is measured perpendicularly to the base.

3. Use volume formulas for cylinders, pyramids, cones, and spheres to solve problems.**

– Sample Problems

(a) Find the volume of the Great Pyramid of Cheops, with square base of side length 754ft, and a height of 482ft.

(b) Find the volume of an equilateral triangular prism with base side length 4cm and height 5cm.

(c) Find the volume of a pyramid with an equilateral triangle base of side length 4cm and height 5cm.

(d) Find the volume of a cylindrical soda can of height 12cm and base diameter 6.5cm. Assuming 1 cubic cm holds 1mL of liquid, and 1mL equals 0.0338 fluid oz, how many ounces would this can hold?

(e) Find the volume of a cone of radius 5 and height 5.

(f) Which volume formulas take the form: volume equals area of base times height? ... volume equals one-third area of base times height?

(g) Suppose that a can of tennis balls is a cylinder that is just large enough to hold three spherical tennis balls. Find the ratio of the total volume of the tennis balls to the total volume of the can if a tennis ball has radius 3cm.

(h) Does your answer to the previous question really depend on the radius of the tennis balls? [What would Archimedes say?]

– Answers to Sample Problems

(a) Find the volume of the Great Pyramid of Cheops, with square base of side length 754ft, and a height of 482ft.

The volume of a pyramid is $\frac{1}{3}Bh$, where B is the area of the base. So,

$$V = \frac{1}{3}(754)^2(482) \approx 91{,}340{,}000 \text{ cubic feet.}$$

(b) Find the volume of an equilateral triangular prism with base side length 4cm and height 5cm.

The volume of a prism is Bh, where B is the area of the base. Using trigonometry (or the 30-60-90 right triangle ratios), we find $B = \frac{\sqrt{3}}{4}s^2$, where s is the side length of the triangle. So, the volume is

$$\frac{\sqrt{3}}{4}(4^2)(5) = 20\sqrt{3} \approx 34.6 \text{ cm}^3.$$

(c) Find the volume of a pyramid with an equilateral triangle base of side length 4cm and height 5cm.

The volume of a pyramid is one-third of its corresponding prism. So the answer to this problem is one-third of the answer to the previous problem: $V = \frac{20\sqrt{3}}{3} \approx 11.5 \text{ cm}^3$.

(d) Find the volume of a cylindrical soda can of height 12cm and base diameter 6.5cm. Assuming 1 cubic cm holds 1mL of liquid, and 1mL equals 0.0338 fluid oz, how many ounces would this can hold?

The radius of the can is 3.25 cm. So the volume is $\pi r^2 h = \pi(3.25)^2(12) \approx 398 \text{ cm}^3$. The can holds $398(0.0338) \approx 13.5$ fluid ounces.

(e) Find the volume of a cone of radius 5 and height 5.

The volume is $\frac{1}{3}\pi r^2 h = \frac{1}{3}\pi(25)(5) \approx 131$ cubic units.

(f) Which volume formulas take the form: volume equals area of base times height? ...volume equals one-third area of base times height?

The volumes of prisms and cylinders are of the form $V = Bh$, where B is the area of the base. The volumes of pyramids and cones are of the form $V = \frac{1}{3}Bh$.

(g) Suppose that a can of tennis balls is a cylinder that is just large enough to hold three spherical tennis balls. Find the ratio of the total volume of the tennis balls to the total volume of the can if a tennis ball has radius 3cm. Notice that the height of the can is 18cm.

$$\begin{aligned} V_{sphs} &= 3\left(\frac{4}{3}\pi \cdot 3^3\right) = 108\pi \text{ cm}^3 \\ V_{cyl} &= \pi(3^2)(18) = 162\pi \text{ cm}^3 \end{aligned}$$

The ratio is therefore $108\pi : 162\pi = 2 : 3$.

(h) Does your answer to the previous question really depend on the radius of the tennis balls? [What would Archimedes say?]
Archimedes knew that this ratios does not depend on the radius of the spheres. To be more general, let's assume a radius r.

$$\begin{aligned} V_{sphs} &= 3\left(\frac{4}{3}\pi r^3\right) = 4\pi r^3 \\ V_{cyl} &= \pi r^2(6r) = 6\pi r^3 \end{aligned}$$

Therefore, in general, the ratio $V_{sphs} : V_{cyl} = 2 : 3$.

- VISUALIZE RELATIONSHIPS BETWEEN TWO-DIMENSIONAL AND THREE-DIMENSIONAL OBJECTS

  – What does it mean for a line to lie in a plane? ...to be perpendicular to a plane? ...to be parallel to a plane?

  A line lies in a plane if every point on the line is also a point in the plane. A line ℓ is perpendicular to a plane P if ℓ intersects P at one point A and if ℓ is perpendicular to any line through point A that lies in plane P. (In advanced math, planes are described by the direction of their perpendicular lines, called "normal vectors.") A line is parallel to a plane if the line and the plane do not intersect.

  – What does it mean for two planes to be parallel? ...perpendicular?

  Two planes are parallel if they do not intersect. Also, two planes are parallel if they are both perpendicular to the same line. Two planes are perpendicular if they intersect at a right angle.

  In other words, suppose plane P is perpendicular to line ℓ and plane Q is perpendicular to line m. Then plane P is parallel to plane Q if lines ℓ and m are parallel, and P is perpendicular to Q if ℓ and m point in perpendicular directions.

  – Are all non-intersecting pairs of lines "parallel?"

  No, just because two lines in space do not intersect does not mean they are "parallel." In space, we have the possibility of "skew" lines. The difference is this: if two lines are parallel, then there exists a plane containing both of them, whereas skew lines lie on parallel planes but never on the same plane. In the picture below, lines k and m are skew. The dashed lines are to help give a three-dimensional perspective.

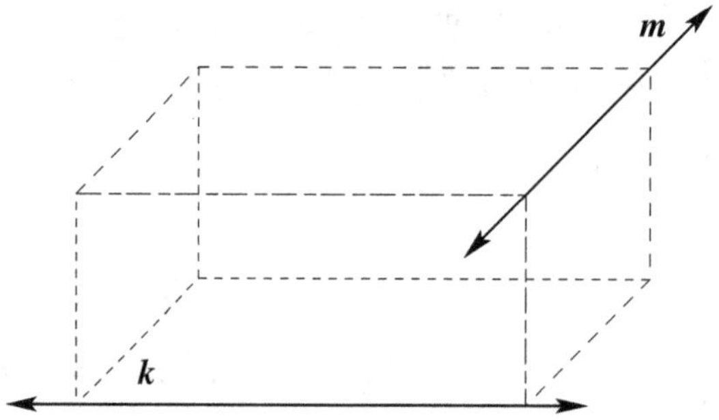

– Why can't skew lines intersect?

Skew lines cannot intersect because they lie on parallel planes, which by definition do not intersect.

– In what ways can two or three lines intersect on a plane?

On a plane, two lines could be parallel, or they could intersect in one point. Three lines could all be parallel, or two could be parallel and one a transversal, or any two of them could intersect in distinct points, or the three lines might be concurrent at a single point.

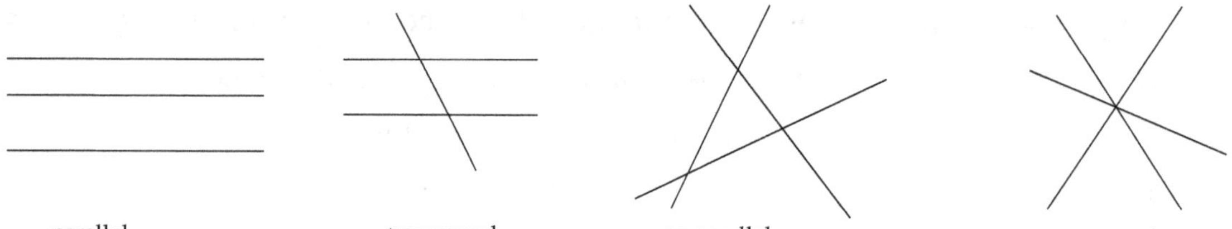

– In what ways can two or three lines intersect in space?

In space, two lines could be parallel or skew, in which case they do not intersect. If they are neither parallel nor skew, then the two lines intersect in one point. Three lines could exhibit the same behavior as in the previous problem, or if two of the lines are skew, then the third line could intersect one of them, both of them, or neither of them.

– In what ways can two or three planes intersect in space?

In space, two planes could be parallel, or else they could intersect in a line. Three planes could all be parallel, or two could be parallel and one could intersect each of the others in a line, or all three planes could intersect in a single point (like two walls and the floor intersect at the corner of a room), or the three planes could intersect in a line, or the three planes could intersect two at a time, with the three resulting lines of intersection being parallel.

- In what ways can a line and a plane intersect in space?

  Either a line is parallel to a plane or else it intersects that plane in a single point.

- How many points determine a line? ... a plane? ... space?

  Two points determine a line. Three non-collinear points determine a plane. Four non-coplanar points determine space.

4. **Identify the shapes of two-dimensional cross-sections of three-dimensional objects, and identify three-dimensional objects generated by rotations of two-dimensional objects.**

   - What do the two-dimensional (2-D) cross-sections of a rectangular prism look like?

     If we restrict ourselves to only those cross-sections that are parallel to one of the sides of the prism, then the only cross-sections we can get are congruent to the faces of the prism.

     vertical slices

     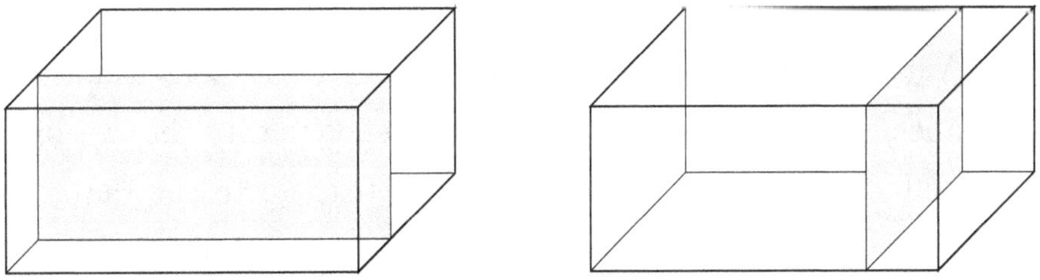

     horizontal slice

     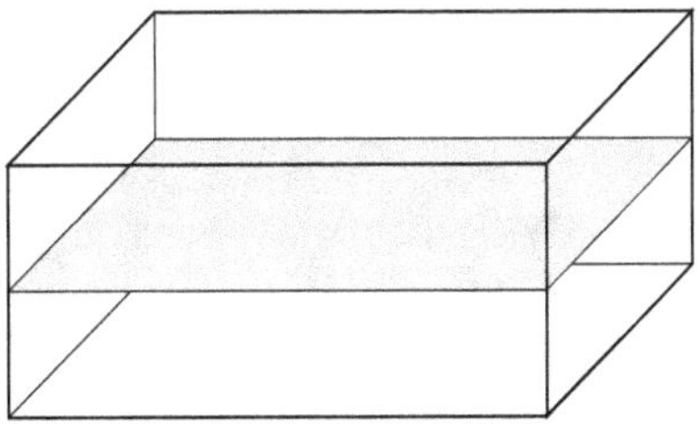

   - What do the 2-D cross-sections of a right pyramid look like?

     If we slice horizontally (parallel to the base), then each cross-section is similar to the base, and possibly degenerate to a point if the slice is tangent to the vertex of the pyramid. Cross-sections perpendicular to the base can be triangles or trapezoids, again possibly degenerate to points or line segments when the slices are tangent to the pyramid.

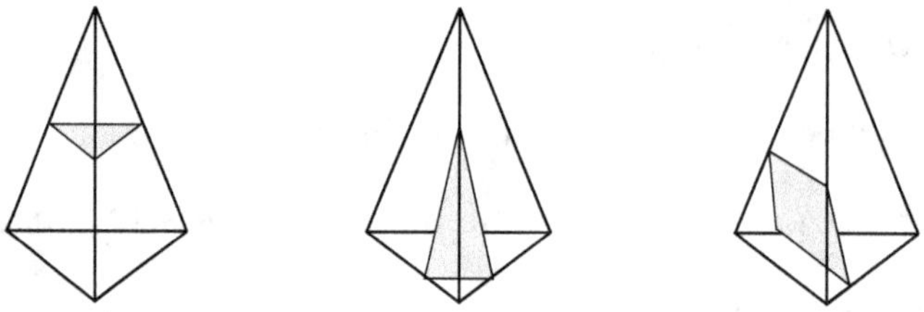

horizontal slice ***vertical slices***

– What do the 2-D cross-sections of a right cylinder look like?
Slices parallel to the base are congruent circles. Slices perpendicular to the base are rectangles of the same height as the cylinder.

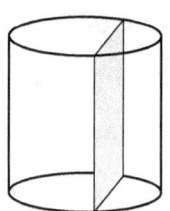

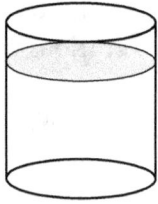

horizontal slice ***vertical slice***

– What do the 2-D cross-sections of a right cone look like?
These cross-sections are so important that they are given the special group name of "conic sections." Circles, ellipses, parabolas, and hyperbolas can all be found in various cross sections of a cone (though you need a double cone to get both branches of the hyperbola). Slices perpendicular to the axis of the cone are circles of different radii, from zero (at the vertex) up to the radius of the base. Slices that are parallel to the slant of the cone are parabolas. Slices at an angle between the slant of the cone and perpendicular to the axis of the cone are ellipses (again, possibly degenerate to a point at the vertex). Other slices are hyperbolas.

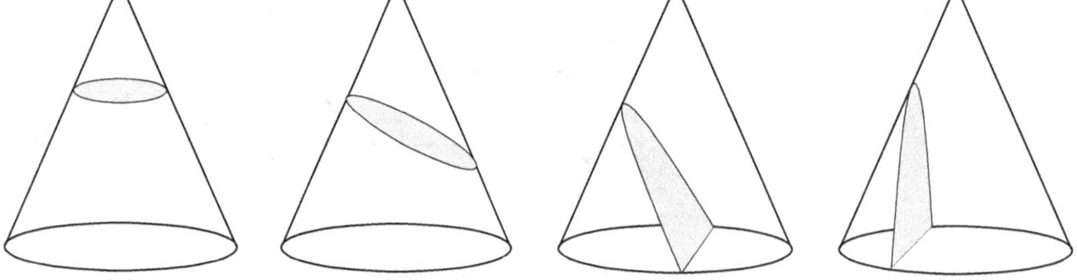

circle ***ellipse*** ***parabola*** ***hyperbola (one branch)***

– What do the 2-D cross-sections of a sphere look like?
A slice of a sphere is a circle of a radius between zero (if the slice is tangent to

the sphere) and the radius of the sphere (e.g., like the Equator on earth).

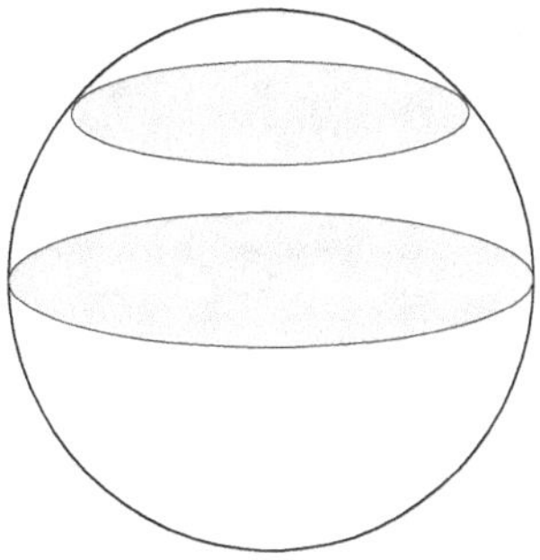

- What three-dimensional (3-D) objects can be formed by rotating 2-D objects?
  Basically, any of the round shapes above – cylinders, cones, and spheres – can be formed by rotating 2-D objects. Cylinders and cones can be formed by rotating a line segment around a line, whereas a sphere can be formed by rotating a circle around one of its diameters.

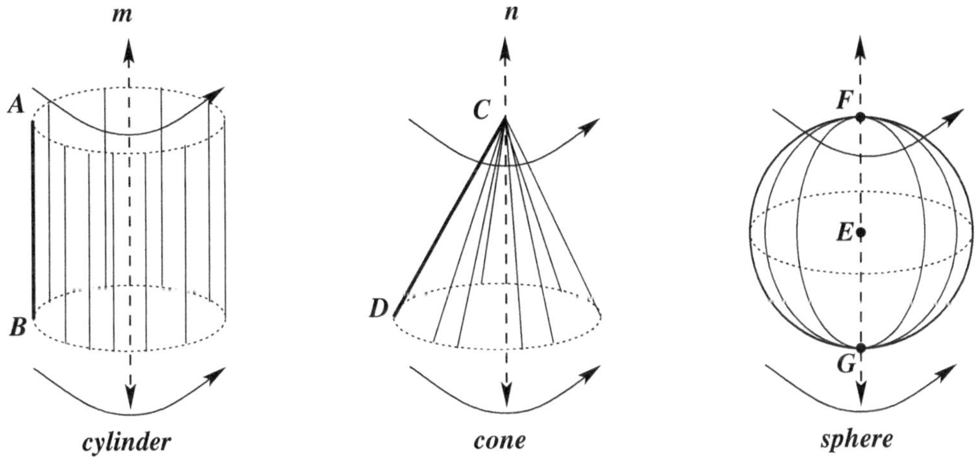

In the picture, rotating $\overline{AB}$ around line m traces out a cylinder; rotating $\overline{CD}$ around n traces out a cone; and rotating circle E around its diameter $\overline{FG}$ traces out a sphere.

- What can happen when you rotate 2-D objects?
  You can get all sorts of shapes by rotating 2-D objects, but they each have an axis of rotational symmetry. Rotating a circle around a tangent line gives a donut-like shape but without a hole. Rotating around a line outside the circle gives a donut-shape, called a "torus."

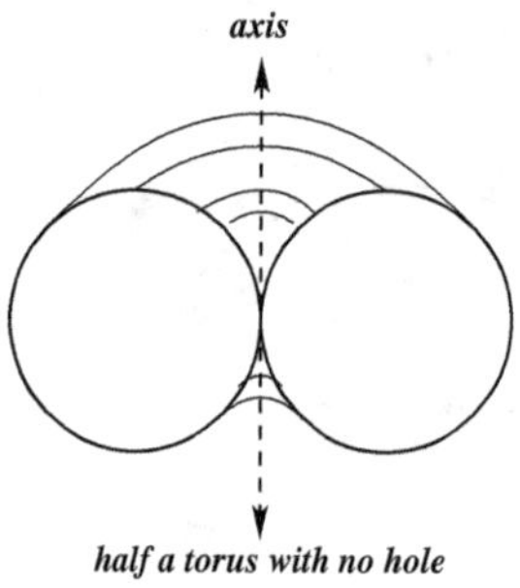

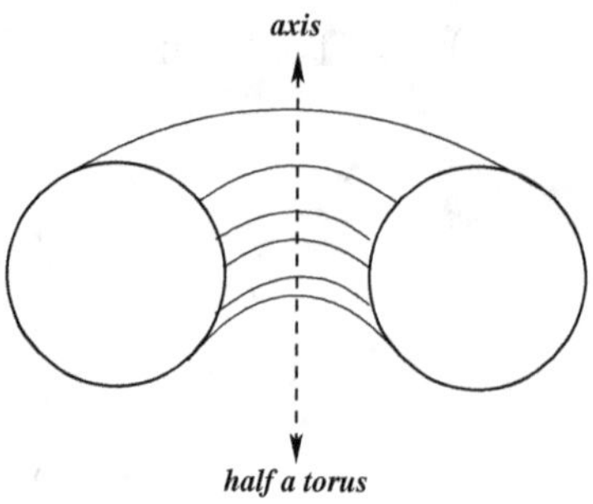

You can even rotate the graph of an arbitrary function $y = f(x)$ around an axis. In the picture below, we rotated some function around the x-axis to generate a solid of revolution. One often studies these solids in integral Calculus.

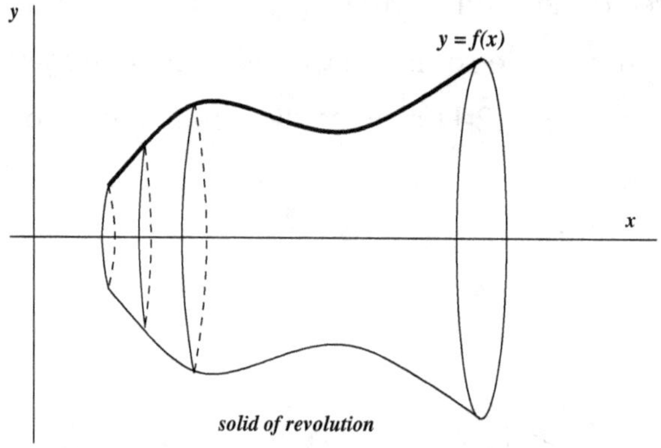

MODELING WITH GEOMETRY

- Apply geometric concepts in modeling situations

1. Use geometric shapes, their measures, and their properties to describe objects (e.g., modeling a tree trunk or a human torso as a cylinder).**

   – What objects could you model as a cylinder or cone (or a frustum of a cone)? How?

   You could definitely model a tree trunk or a human torso as a cylinder. You could model an ice cream cone as a cone, of course, and maybe even a tree trunk as a cone, too, if you take the tree trunk all the way to the top of the tree. A frustum of a cone is the shape that you would add to the bottom of a cone to make it into a bigger cone. Put another way, it's the region that remains if you slice a cone horizontally and remove the cone-shaped piece. It's like a cylinder, except the radius at the top is different than the radius at the bottom. So a frustum might be a better model for a tree or a torso, but it is certainly more complicated.

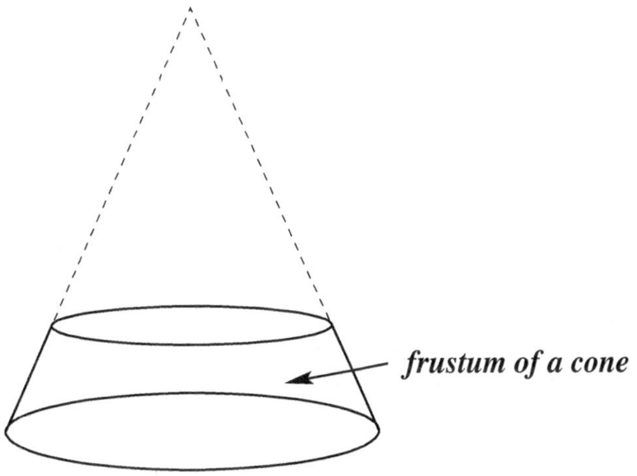

   Suppose you want to estimate how much wood is in the body of a tree (not including the branches), and you know that the height of the tree is 30 feet, and that the circumference of the trunk is 7 feet. You also notice that the trunk looks pretty cylindrical for the first ten feet of height, and then it starts to taper inward until it reaches the top.

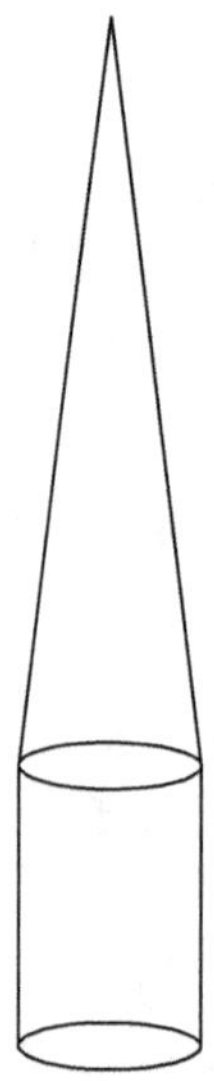

So, you decide to model the tree as a cylinder with a cone on top of it. The cylinder has height 10 feet and circumference 7 feet, which means that the radius is $\frac{7}{2\pi} \approx 1.1$ feet. The cone has a height of $30 - 10 = 20$ feet and the same radius as the cylinder. So the total amount of wood would be the total volume, which equals the volume of the cylinder plus the volume of the cone:

$$V_{\text{cyl}} + V_{\text{cone}} \approx \pi \left(\frac{7}{2\pi}\right)^2 (10) + \frac{1}{3}\pi \left(\frac{7}{2\pi}\right)^2 (20) \approx 64.99 \text{ cubic feet.}$$

– What objects could you model as a prism or a pyramid (or a frustum)? How? You could model shipping boxes, boxcars, and desktop nameplates as prisms. The famous pyramids of Egypt make great pyramidal models. The simplest Platonic solid, the tetrahedron, is a pyramid with four equilateral triangles as its four sides. Similar to a frustum of a cone, a frustum of a pyramid is the part that remains if you remove a pyramid from a larger pyramid. It is like a prism, but with a top that is smaller than the base.

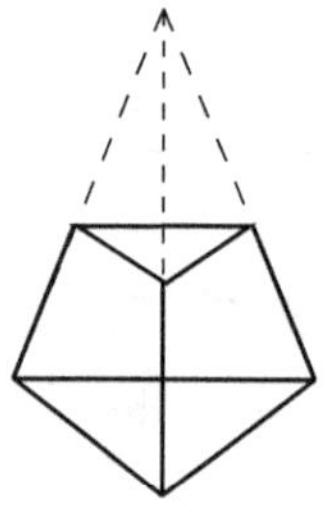

frustum of a pyramid

For several modeling problems, see the Sample Problems on page 109.

– What objects could you model as a sphere? How?

In astronomy, planets, moons, and stars are often modeled as spheres, even though they're not exactly spherical. For instance, the radius of the Earth varies a little, but is about 6378 km. So the volume of the Earth is about $\frac{4}{3}\pi(6378)^3 \approx 1.09 \times 10^{12}$ cubic kilometers.

On a smaller scale, balls used for sports can often be modeled as spheres. A golf ball has a radius of about 0.84 inches. So its volume would be $\frac{4}{3}\pi(0.84)^3 \approx 2.48$ cubic inches.

2. Apply concepts of density based on area and volume in modeling situations (e.g., persons per square mile, BTUs per cubic foot).**

   – How can you use population density to determine population?

     Consider the units. If the population density of a rectangular town averages 24 people per square mile, and the area of the town is 100 square miles, then there are $(24)(100) = 2400$ people in the town. You can see this using the unit labels, provided you remember that "per" comes before a label in the denominator. So the previous problem could be written:

$$\left(\frac{24 \text{ people}}{\text{sq mile}}\right)(100 \text{ sq mile}) = 2400 \text{ people},$$

     because the units "sq miles" cancel out.

   – How can you use density to determine mass?

     Here, density is often given as a measure of mass per unit volume. For instance, the density of aluminum is 2.8 grams per cubic centimeter. So, if you had 36 cubic centimeters of aluminum, the total mass would be:

$$\left(\frac{2.8 \text{ g}}{\text{cm}^3}\right)(36 \text{ cm}^3) = 100.8 \text{ g}.$$

     So the total mass was equal to the density times the volume. This makes sense because here, density is mass divided by volume.

   – Is there a general principle at work here? If so, what is it?

     There is definitely a general principle at work here, and it eventually reaches all the way to calculus and beyond. It is the idea of a "rate" or a ratio. Here, we have seen a ratio of two different quantities, like people and square miles, and we used a known relationship between them in a useful way. We could generalize these density statements as something like:

$$(\text{Density of stuff per unit of space}) \times (\text{total space}) = \text{total amount of stuff}.$$

In calculus, one often works with rates of change and total change of various quantities.

3. Apply geometric methods to solve design problems (e.g., designing an object or structure to satisfy physical constraints or minimize cost; working with typographic grid systems based on ratios).**

   – What is Passive Solar Design? How can we use geometry to model it?

   Passive Solar Design is the idea of using sunlight to help heat your home in the winter, but keeping direct sunlight out in the summer. It takes advantage of the fact that the sun doesn't go as high in the sky in winter as it does in the summer in many parts of the world. In the Northern Hemisphere, by building an overhang above a southern window, you can let in sunlight in winter, but shade the window in summer.

   For instance, according to a website[5], the sun in Phoenix, Arizona, reaches a maximum angle of about 80 degrees at the summer solstice and about 33 degrees at the winter solstice. That's quite a range!

   Suppose we want to design a window that is in full sun at the winter solstice and fully shaded at the summer solstice in Phoenix. Consider the picture below. Suppose the overhang is one foot in length. Find y and w to the nearest inch so that the window lets in full sun at the winter solstice, but is fully shaded at the summer solstice. Angle measures are given in degrees.

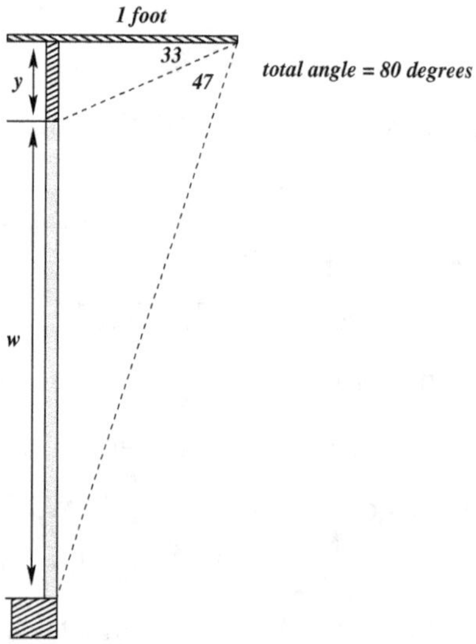

Notice that we have two right triangles, with right angles where the overhang

[5]http://www.timeanddate.com/worldclock/astronomy.html?n=197&month=6&year=2008&obj=sun&afl=-11&day=1, accessed August 2012.

meets the wall. To find y, we need to solve: $\tan 33 = \frac{y}{1}$. Using a calculator, we get $y \approx 0.649$ feet, or about 7.8 inches. Then, to find w, we use another tangent relationship:

$$\tan 80 = \frac{w+y}{1},$$

from which we get $w \approx 5.022$ feet, or 5 feet and 0.26 inches.

– How does font size relate to ratios? How does the amount of ink needed to print a document relate to ratios?

Font size is based on points, and 72 points make up one inch. So, a 12-point font has a height that is one-sixth of an inch. What if you start with a 36-point font, and then scale it up to a 72-point font? Well, you have doubled the size of the font. That is like performing a dilation with a scale factor of 2. That means the heights of the letters would double and the width of each letter would double as well.

What happens to the amount of ink needed? Well, the ink needed to make a letter could be thought of as the area that letter takes up on the page. If you double the length and width of a letter, then you have multiplied its area by 4. So you would need four times as much ink whenever you double the font size, assuming of course that a 72-point font is exactly a scaled-up version of a 36 point font. (Warning: picture may not be exactly to scale.)

36–point font ***72–point font***

So, in general, if your font size goes up by a factor of k, then your letters would get k times wider and k times taller. The amount of ink needed would go up by a factor of k^2.

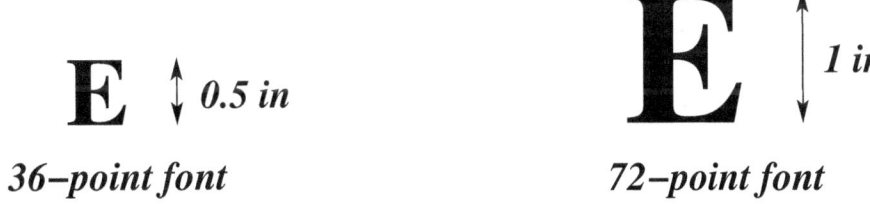